Variational Calculus with Engineering Applications

Variational Calculus with Engineering Applications

Constantin Udriste and Ionel Tevy

Registered Office(s)
John Wiley & Sons, Inc., 111 River Street, Hoboken, NJ 07030, USA
John Wiley & Sons Ltd, The Atrium, Southern Gate, Chichester, West Sussex, PO19 8SQ, UK

Editorial Office(s)
9600 Garsington Road, Oxford, OX4 2DQ, UK
The Atrium, Southern Gate, Chichester, West Sussex, PO19 8SQ, UK

For details of our global editorial offices, customer services, and more information about Wiley products visit us at www.wiley.com.

Wiley also publishes its books in a variety of electronic formats and by print-on-demand. Some content that appears in standard print versions of this book may not be available in other formats.

A catalogue record for this book is available from the Library of Congress

Hardback ISBN: 9781119944362; ePDF ISBN: 9781119944379; epub ISBN: 9781119944386; oBook ISBN: 9781119944423

Cover image: © enjoynz/Getty Images
Cover design by Wiley

Set in 9.5/12.5pt STIXTwoText by Integra Software Services Pvt. Ltd, Pondicherry, India
Printed and bound by CPI Group (UK) Ltd, Croydon, CR0 4YY

C9781119944362_101022

Contents

Preface

The Variational Calculus with Engineering Applications was and is being taught to fourth-year engineering students, in the Faculty of Applied Sciences, Mathematics - Informatics Department, from the University Politehnica of Bucharest, by Prof. Emeritus Dr. Constantin Udriste. Certain topics are taught at other faculties of our university, especially as master's or doctoral courses, being present in the papers that can be published in journals now categorized as "ISI".

The chapters were structured according to the importance, accessibility, and impact of the theoretical notions able to outline a future specialist based on mathematical optimization tools. The probing and intermediate variants lasted for a number of sixteen years, leading to the selection of the most important manageable notions and reaching maturity through this variant that we decided to publish at Wiley.

Now the topics of the book include seven chapters: Extrema of Differentiable Functionals; Variational Principles; Optimal Models Based on Energies; Variational Integrators; Miscellaneous Topics; Extremals with Nonholonomic Constraints; Problems: Free and Constrained Extremals. To cover modern problem-solving methods, each chapter includes Maple application topics. The scientific authority we have as professors in technical university has allowed us to impose our point of view on the subject and on the types of reasoning that deserve to be offered to readers of variational calculus texts. Everything has been structured for the benefit of functional optimization that describes significant physical or engineering phenomena. To unlock the students' minds we preferred a simplified language, specific to applied mathematics, sending readers of pure mathematics to the bibliography. Long experience as mathematics teachers in technical universities has prepared us to fluently expose ideas undressed by excessive formalizations, with great impact on the students' understanding of the natural phenomena encountered in engineering and economics as well as their dressing in mathematical clothing. In this sense, we preferred mathematical modeling as a scientific support for intelligent presentation of the real world and we avoided the hint of abstract notions specific to pure mathematics.

Topics that distinguish this book from existing books include: Maple application topics, invexity tests, case of path-independent curvilinear integral functionals, variational integrators, discretization of Hamilton ODEs, numerical study of vibrating string motion, numerical study of vibrating membrane motion, linearization of nonlinear ODEs and

PDEs, geometric dynamics and wind, extremals with nonholonomic constraints, free and constrained extremals. In fact, this is the first book that studies the curvilinear integral functionals at the level of student courses. The novelty issues were finalized following repeated discussions with the professors from our faculty and from Faculty of Mathematics in West University of Timisoara, especially with those interested in functional optimization, to whom we bring thanks and thoughts of appreciation. We are open to any comments or criticisms which bring didactic benefits for variational calculus theories.

The mottos are stanzas from two poems by Tudor Arghezi (1880–1967), a Romanian writer, known for his contribution to the development of Romanian lyric poetry under the influence of Baudelaireanism. His poetic work, of exemplary originality, represents another significant age of Romanian literature.

Born in Bucharest of peasant stock, Tudor Arghezi was awarded Romania's National Poetry Prize in 1946 and the State Prize for Poetry in 1956.

The authors thank Associate Prof. Dr. Oana-Maria Pastae, "Constantin Brancusi" University of Tg-Jiu, for the English improvement of the manuscript. We are indebted to Prof. Dr. Dumitru Opris from the West University of Timisoara for the solutions offered to some discretization problems.

Bucharest, April 2022

1

Extrema of Differentiable Functionals

Motto:
"I wrote them with my nail on the plaster
On a wall of empty cracks,
In the dark, in my solitude,
Unaided by the bull lion vulture
of Luke, Mark and John."
Tudor Arghezi – *Flowers of Mildew*

Variational calculus aims to generalize the constructions of classical analysis, to solve extrema problems for functionals. In this introductory chapter, we will study the problem of differentiable functionals defined on various classes of functions. The news refers to path integral functionals. For details, see [4, 6, 7, 8, 14, 18, 21, 30, 58].

1.1 Differentiable Functionals

Let us consider the point $x = (x^1, x^2, ..., x^n)$ and the function $f : D \subset \mathbb{R}^n \to \mathbb{R}, x \mapsto f(x)$. If this function has continuous partial derivatives with respect to each of the variables $x^1, x^2, ..., x^n$, then increasing the argument with $h = (h^1, h^2, ..., h^n)$ produces

$$f(x + h) - f(x) = \sum_{j=1}^{n} \frac{\partial f}{\partial x^j}(x) h^j + r(x, h).$$

The first term on the right-hand side represents the differential of the function f at the point x, a linear form of argument growth, the vector h; the second term is the deviation from the linear approximation and is small in relation to h, in the sense that

$$\lim_{h \to 0} \frac{r(x, h)}{||h||} = 0.$$

Let $\mathbf{U}, \mathbf{V}$ be two normed vector spaces. The previous definition can be extended immediately to the case of functions $f : D \subset \mathbf{U} \to \mathbf{V}$. If $\mathbf{U}$ is a space of functions, and $\mathbf{V} = \mathbb{R}$, then instead of function we use the term *functional*.

Variational Calculus with Engineering Applications, First Edition. Constantin Udriste and Ionel Tevy.
© 2023 John Wiley & Sons Ltd. Published 2023 by John Wiley & Sons Ltd.

Definition 1.1: The function f is called differentiable at a point $x \in D$ if there exist a linear and continuous operator $\mathbf{d}f(x, \cdot)$ and a continuous function $r(x, \cdot)$ such that for any vector $h \in \mathbf{U}$ to have

$$1^0 \qquad f(x + h) - f(x) = \mathbf{d}f(x, h) + r(x, h),$$

$$2^0 \qquad \lim_{h \to 0} \frac{r(x, h)}{\|h\|} = 0.$$

The linear continuous operator $\mathbf{d}f(x, \cdot)$ is called the *derivative* of the function f at given point x.

For a given nonzero vector $h \in \mathbf{U}$ and $t > 0$, the vector

$$\delta(x, h) = \lim_{t \to 0} \frac{f(x + th) - f(x)}{t} \in \mathbf{V},$$

if the limit exists, is called the *variation or derivative of the function f at the point x along the direction h.*

A differentiable function of real variables has a derivative along any direction. The property is also kept for functions between normed vector spaces. Specifically, the next one takes place:

Proposition 1.1: *If the function $f : D \subset \mathbf{U} \to \mathbf{V}$ is differentiable at the point $x \in D$, then for any nonzero vector $h \in \mathbf{U}$ the function $f(x + th)$, of real variable $t \geq 0$, is derivable with respect to t, for $t = 0$ and*

$$\frac{df}{dt}(x + th)\Big|_{t=0} = \delta f(x, h) = \mathbf{d}f(x, h).$$

Proof. The derivative sought is obviously $\delta f(x, h)$ and then

$$\frac{df}{dt}(x + th)\Big|_{t=0} = \delta f(x, h) = \lim_{t \to 0} \frac{f(x + th) - f(x)}{t}$$

$$= \lim_{t \to 0} \frac{\mathbf{d}f(x, th) + r(x, th)}{t} = \lim_{t \to 0} \frac{t\mathbf{d}f(x, h) + r(x, th)}{t}$$

$$= \mathbf{d}f(x, h) + \lim_{th \to 0} \frac{r(x, th)}{\|th\|} \|h\| = \mathbf{d}f(x, h). \qquad \square$$

Example 1.1: *Any linear continuous operator $T : \mathbf{U} \to \mathbf{V}$ is, obviously, a differentiable function at any point x and $\delta T(x, h) = \mathbf{d}T(x, h) = T(h)$ since*

$$T(x + h) - T(x) = T(h).$$

Let f be a functional. The simplest examples of *functionals* are given by formulas: (i) evaluation functional ("application of a function on the value at a point"), $f \mapsto f(x_0)$, where x_0 is a fixed point; (ii) definite integration (functional defined by definite integral); (iii) numerical quadrature defined by definite integration:

$$I(f) = a_0 f(x_0) + a_1 f(x_1) + \dots + a_n f(x_n) = \int_a^b f(x)dx, \ \forall f \in P_n,$$

where P_n means the set of all polynomials of degree at most n; and (iv) distributions in analysis (as linear functionals defined on spaces of test functions). We will continue to deal with definite integral type functionals, as the reasoning is more favorable to us.

Example 1.2: *Let us consider the functional*

$$J(x(\cdot)) = \int_a^b L(t, x(t))dt,$$

defined on the space $C^0[a, b]$ of the continuous functions on the segment $[a, b]$, endowed with the norm of uniform convergence. The Lagrangian of the functional (i.e., the function $L(t, x)$) is presumed continuous and with continuous partial derivatives of the first order in the domain $a \leq t \leq b$, $-\infty < x < +\infty$.

Let us determine the variation of the functional $J(x(\cdot))$ when the argument $x(t)$ increases with $h(t)$:

$$\triangle J(x(\cdot)) = J(x + h) - J(x) = \int_a^b [L(t, x(t) + h(t)) - L(t, x(t))]dt.$$

Since the Lagrangian L is a differentiable function, we have

$$L(t, x + h) - L(t, x) = \frac{\partial L}{\partial x} h + r(t, x, h),$$

where

$$\lim_{h \to 0} \frac{r(t, x, h)}{h} = 0.$$

Therefore

$$\delta J(x, h) = \int_a^b \frac{\partial L(t, x(t))}{\partial x} h(t)dt.$$

In this way, according to the previous proposition and the derivation formula for integrals with parameters, we find

$$\delta J(x, h) = \frac{dJ}{d\varepsilon}(x + \varepsilon h)\Big|_{\varepsilon=0} = \frac{d}{d\varepsilon} \int_a^b L(t, x(t) + \varepsilon h(t))dt\Big|_{\varepsilon=0}$$

$$= \int_a^b \frac{\partial L(t, x(t))}{\partial x} h(t)dt. \tag{1}$$

Example 1.3: *Let us consider the functional*

$$J(x(\cdot)) = \int_a^b L(t, x(t), x'(t))dt,$$

defined on the space $C^1[a, b]$ of functions with a continuous derivative on the segment $[a, b]$, endowed with the norm of uniform convergence of derivatives. The Lagrangian $L(t, u, v)$ of the functional is supposed to have first-order continuous partial derivatives. To write the integral functional, we use in fact the pullback form $L(t, x(t), x'(t))$ of the Lagrangian.

Applying the derivation formula of the integrals with parameter we obtain

$$\delta J(x, h) = \frac{dJ}{d\varepsilon}(x + \varepsilon h)\Big|_{\varepsilon=0}$$

$$= \frac{d}{d\varepsilon} \int_a^b L(t, x(t) + \varepsilon h(t), x'(t) + \varepsilon h'(t))dt\Big|_{\varepsilon=0}$$

$$= \int_a^b \left[\frac{\partial L}{\partial x}(t, x(t), x'(t)) h(t) + \frac{\partial L}{\partial x'}(t, x(t), x'(t)) h'(t) \right] dt. \tag{2}$$

Analogously, we can extend the result to the functional

$$J(x(\cdot)) = \int_a^b L(t, x(t), x^{(1)}(t), ..., x^{(m)}(t)) \, dt,$$

defined on the space $C^m[a, b]$ of functions with continuous derivatives up to the order m inclusive, on the segment $[a, b]$. In this way we have

$$\delta J(x, h) = \int_a^b \left[\frac{\partial L}{\partial x} h(t) + \frac{\partial L}{\partial x^{(1)}} h^{(1)}(t) + ... + \frac{\partial L}{\partial x^{(m)}} h^{(m)}(t) \right] dt. \tag{3}$$

Example 1.4: *Let us consider a functional which depends on several function variables, for example*

$$J(x^1(\cdot), ..., x^n(\cdot)) = \int_a^b L(t, x^1(t), ..., x^n(t), \dot{x}^1(t), ..., \dot{x}^n(t)) \, dt,$$

defined on the space $(C^1[a, b])^n$, whose elements are vector functions $x(t) = (x^1(t), ..., x^n(t)), t \in [a, b]$, with the norm of uniform convergence of derivatives

$$\|x\| = \max_{t\in[a,b]} [|x^1(t)|, ..., |x^n(t)|, |\dot{x}^1(t)|, ..., |\dot{x}^n(t)|]$$

($\dot{x}$ means the derivative with respect to t).

Supposing that the Lagrangian L has continuous partial derivatives, denoting $h(t) = (h^1(t), ..., h^n(t))$, the variation of the functional J is

$$\delta J(x, h) = \int_a^b \left[\frac{\partial L}{\partial x^i} h^i(t) + \frac{\partial L}{\partial \dot{x}^i} \dot{h}^i(t) \right] dt, \ i = \overline{1, n} \tag{4}$$

(sum over the index i).

Now, let us consider functionals whose arguments are functions of several real variables.

Example 1.5: *Let $\Omega \subset \mathbb{R}^2$ be a compact domain. As an example we will take, first, the double integral functional*

$$J(w(\cdot)) = \iint_\Omega L(x, y, w(x, y), w_x(x, y), w_y(x, y)) \, dx \, dy,$$

where we noted for abbreviation

$$w_x = \frac{\partial w}{\partial x}, \quad w_y = \frac{\partial w}{\partial y}.$$

The functional is defined on the space $C^1(\Omega)$ of all functions with continuous partial derivatives; the norm of the space $C^1(\Omega)$ is given by

$$\|w\| = \max_{(x,y)\in\Omega} [\, |w(x, y)|, |w_x(x, y)|, |w_y(x, y)| \,].$$

Supposing that the Lagrangian L has continuous partial derivatives, the variation of the functional J, as the argument w grows with $h(x, y)$, is

$$\delta J(w, h) = \frac{dJ}{d\varepsilon}(w + \varepsilon h)\Big|_{\varepsilon=0}$$

$$= \frac{d}{d\varepsilon} \iint_\Omega L(x, y, w + \varepsilon h, w_x + \varepsilon h_x, w_y + \varepsilon h_y) \, dxdy\Big|_{\varepsilon=0}$$

$$= \iint_\Omega \left(\frac{\partial L}{\partial w} h + \frac{\partial L}{\partial w_x} h_x + \frac{\partial L}{\partial w_y} h_y \right) dx \, dy. \tag{5}$$

Example 1.6: *Another example is the curvilinear integral functional*

$$J(w(\cdot)) = \int_\Gamma [L_1(x, y, w(x, y), w_x(x, y), w_y(x, y)) \, dx$$

$$+ L_2(x, y, w(x, y), w_x(x, y), w_y(x, y)) \, dy \,],$$

where Γ is a piecewise C^1 curve which joins two fixed points A, B in a compact domain $\Omega \subset \mathbb{R}^2$.

Suppose the argument $w(x, y)$ grows with $h(x, y)$. Then the variation of the functional is

$$\delta J(w, h) = \int_\Gamma \left(\frac{\partial L_1}{\partial w} h + \frac{\partial L_1}{\partial w_x} h_x + \frac{\partial L_1}{\partial w_y} h_y \right) dx$$

$$+ \left(\frac{\partial L_2}{\partial w} h + \frac{\partial L_2}{\partial w_x} h_x + \frac{\partial L_2}{\partial w_y} h_y \right) dy. \tag{6}$$

1.2   Extrema of Differentiable Functionals

Let us consider the functional $J : D \subset \mathbf{U} \to \mathbb{R}$, defined on a subset D of a normed vector space $\mathbf{U}$ of functions. By definition, the functional J has a *(relative) minimum* at the point x_0 in D, if there exists a neighborhood V, of point x_0, such that

$$J(x_0) \leq J(x), \quad \forall x \in V \cap D.$$

If the point x_0 has a neighborhood V on which the opposite inequality takes place

$$J(x_0) \geq J(x), \quad \forall x \in V \cap D,$$

we say that x_0 is a point of *local maximum* for the functional J. The minimum and maximum points are called *relative extrema* points.

In classical analysis, the extrema points of a differentiable function are among the critical points, that is, among the points that cancels first-order derivatives. A similar property occurs in the case of functional ones on normed vector spaces of functions, only in this case the extrema points are found between the extremals (critical points).

Proposition 1.2: *If the function x is an extremum point for the functional J, interior point of the set D and if J is differentiable at this point, then $\delta J(x, h) = 0$ for any growth h.*

Proof.   Let h be a growth of the argument x (function); since x is an interior point of the set D, the function $J(x + \varepsilon h)$ of real variable ε is defined on hole interval $[-1, 1]$. This function has an extremum point at $\varepsilon = 0$ and is derivable at this point. Then its derivative must vanish at $\varepsilon = 0$. It follows

$$\delta J(x, h) = \frac{dJ}{dt}(x + th)\Big|_{t=0} = 0.$$

□

Any point x at which the variation $\delta J(x, h)$ of the functional J is canceled identically with respect to h is called either the *stationary point* or *critical point* or *extremal* of the functional.

Hence, *for establishing the extrema points of a functional, the variation $\delta J(x, h)$ must be expressed, determine the critical points (those at which the variation is canceled identically with respect to h) and then choose from these the minimum points or the maximum points.*

Commentary The variation $\delta J(x, h)$ of the functional J is a continuous linear functional on the normed vector space $\mathbf{U}$ on which is defined J. The set $\mathbf{U}^*$ of all continuous linear functionals on $\mathbf{U}$ is called *dual space*. Thus the evaluation of the variation of a functional imposes the existence of some representation theorems for the dual space.

For example, for any linear functional $T : \mathbb{R}^n \to \mathbb{R}$ there exists a vector $a = (a^1, ..., a^n) \in \mathbb{R}^n$ such that

$$T(h) = a^1 h^1 + ... + a^n h^n = \langle a, h \rangle, \quad \forall h \in \mathbb{R}^n.$$

Also, any continuous and linear functional $T : C^0[a, b] \to \mathbb{R}$, on the space of continuous functions, can be written in the form

$$T(h) = \int_a^b h(t) \, dg_T(t),$$

where g_T is a function with bounded variation and continuous at the right (*F. Riesz's Theorem*).

To determine the extrema points, we need the identical cancelation of the first variation. For that we use

Lemma 1.1: **(Fundamental lemma of variational calculus)** *Let x be a real continuous function on the interval $[a, b]$. If for any continuous function h, vanishing at a and b, the equality*

$$\int_a^b x(t)h(t)dt = 0,$$

is true, then $x(t) = 0, \ \forall t \in [a, b]$.

Proof. Suppose there exists a point $t_0 \in [a, b]$ at which $x(t_0) \neq 0$. Let $x(t_0) > 0$. Then there exists a neighborhood $V = (t_0 - \varepsilon, t_0 + \varepsilon)$ of t_0 where $x(t) > 0$. Let us consider a continuous function h, strictly positive on the neighborhood V and null outside it (for example, $h(t) = (t - t_0 + \varepsilon)(t_0 + \varepsilon - t)$ for $t \in V$). Then

$$\int_a^b x(t)h(t) \, dt = \int_{t_0 - \varepsilon}^{t_0 + \varepsilon} x(t)h(t) \, dt > 0,$$

which contradicts the hypothesis and concludes the proof of the lemma. $\qquad\square$

Remark 1.1: 1) For the set of continuous real functions $C^0[a, b]$, the integral $\langle x, y \rangle = \int_a^b x(t)h(t) \, dt$ is a scalar product. Consequently,

$$\langle x, y \rangle = \int_a^b x(t)h(t) \, dt = 0, \forall h(t)$$

implies $x(t) = 0$.

2) The set of functions h you need for checking the equality in the lemma can often be reduced. As can be seen from the proof of the fundamental lemma, we can impose derivability conditions of any order to the functions h and cancelation at the end of the interval etc. It is essential that for any $t \in [a, b]$ and any $\varepsilon > 0$ to exist, in the set of test functions, a strictly positive function on the interval $(t_0 - \varepsilon, t_0 + \varepsilon)$ and null function otherwise.

3) In a similar wording, the fundamental lemma of the variational calculus remains in force for functions of several variables.

4) We can also see that in previous constructions it is sufficient that the functional J be defined on a linear variety (affine subspace) or at least on a convex set: with two points x and $x + h$ to be in the domain of J all points of the form $x + th$ for any $t \in \mathbb{R}$ or at least $t \in [-1, 1]$ too. Specifically, we can impose restrictions (conditions) on extrema points and variations h, but remaining in a linear variety.

Problem 1.2.1: **a)** Let x be a C^k function on the interval $[a, b]$. If for any $h \in C^k[a, b]$, null at a and b, together with its first $k - 1$ derivatives, the equality

$$\int_a^b x(t)\, h^{(k)}(t)\, dt = 0$$

holds, then x is a polynomial of $deg \leq k - 1$.

b) The previous statement remains in force even if x is only a continuous function.

Solution. **a)** The property "x is a polynomial of $deg \leq k - 1$" is equivalent to "$x^{(k)} = 0$, for any $t \in [a, b]$". We are inspired by the proof of the fundamental lemma. Suppose there exists a point $t_0 \in [a, b]$ at which $x^{(k)}(t_0) \neq 0$. Let $x^{(k)}(t_0) > 0$. Then there exists a neighborhood $V = (t_0 - \varepsilon, t_0 + \varepsilon)$, of t_0 where $x^{(k)}(t) > 0$. Let us consider the C^k function $h(t) = (t - t_0 + \varepsilon)^k (t_0 + \varepsilon - t)^k$ for $t \in V$ and null outside V. Assuming that $V \subset (a, b)$, the function h, together with its first $k - 1$ derivatives vanishes in a and b. Integrating k times by parts one obtains

$$\int_a^b x(t) h^{(k)}(t)\, dt = x(t) h^{(k-1)}(t)|_a^b - \int_a^b x'(t) h^{k-1}(t)\, dt = \ldots$$

$$= (-1)^k \int_a^b x^{(k)}(t) h(t)\, dt = (-1)^k \int_{t_0 - \varepsilon}^{t_0 + \varepsilon} x^{(k)}(t) h(t)\, dt \neq 0,$$

(> 0 or < 0 as k is even or odd) which contradicts the hypothesis.

b) Firstly, let us prove the following result:

A function g, continuous on $[a, b]$ is the order k derivative of a C^k function h which, together with its first $k - 1$ derivatives, vanishes in a and b, $g = h^{(k)}$, if and only if

$$\int_a^b t^p g(t)\, dt = 0, \quad \forall p \in \{0, 1, \ldots, k - 1\}. \tag{P}$$

Obviously, if $g = h^{(k)}$, the property (P) results by integrating by parts as many times as necessary.

Conversely, let us denote

$$h(t) = \int_a^t \int_a^{s_k} \ldots \int_a^{s_2} g(s_1)\, ds_1 \ldots ds_k = \frac{1}{n!} \int_a^t g(s)(t - s)^{k-1}\, ds.$$

Then $h^{(k)} = g$ and $h(a) = h'(a) = \ldots h^{(k-1)}(a) = 0$, and the equalities in the property (P), in which we put $g = h^{(k)}$ and integrate by parts, give us successively $h^{(k-1)}(b) = \ldots = h'(b) = h(b) = 0$.

Returning to the proposed problem, let us note

$$g(t) = x(t) - (c_0 + c_1 t + ... + c_{k-1} t^{k-1})$$

and determine the coefficients $c_0, c_1, ...c_{k-1}$ so that the function g verifies the equalities of the property (P).

The orthogonalization properties in a Euclidean vector space ensure the possibility of this determination. Then we can take $h^{(k)} = g$ and get

$$0 = \int_a^b x(t) g(t) \, dt = \int_a^b (g(t) + (c_0 + c_1 t + ... + c_{k-1} t^{k-1})) g(t) \, dt$$

$$= \int_a^b g^2(t) \, dt + \int_a^b (c_0 + c_1 t + ... + c_{k-1} t^{k-1}) g(t) \, dt = \int_a^b g^2(t) \, dt.$$

The result is $g(t) = 0$, $\forall t \in [a, b]$, i.e., x is a polynomial of degree at most $k - 1$.

Remark 1.2: The previous result illustrates on a concrete situation the equality $U^{\perp\perp} = U$, for any closed subspace of a Hilbert space.

Let us examine the detection of critical points (extremals) for the functionals whose variations were determined in the previous section.

Example 1.7: *Obviously, a nonzero continuous linear functional* $T : \mathbf{U} \to \mathbb{R}$ *has no critical points.*

Example 1.8: *For the functionals of type*

$$J(x(\cdot)) = \int_a^b L(t, x(t)) \, dt,$$

the variation is given by formula (1). Then, applying the fundamental lemma we obtain

$$\frac{\partial L}{\partial x}(t, x(t)) = 0. \tag{7}$$

Equation (7) can be solved, according to the implicit function theorem, obtaining solutions $x = x(t)$, *around the points* (t_0, x_0) *at which*

$$\frac{\partial L}{\partial x}(t_0, x_0) = 0, \quad \frac{\partial^2 L}{\partial x^2}(t_0, x_0) \neq 0.$$

Example 1.9: *A special place, generic in the variational calculus, has the functional*

$$J(x(\cdot)) = \int_a^b L(t, x(t), x'(t)) \, dt,$$

where L is a C^2 Lagrangian. The variation of this functional is given in formula (2):

$$\delta J(x, h) = \int_a^b \left[\frac{\partial L}{\partial x}(t, x(t), x'(t)) h(t) + \frac{\partial L}{\partial x'}(t, x(t), x'(t)) h'(t) \right] dt.$$

Integrating the second term by parts, the variation takes the shape

$$\delta J(x,h) = \int_a^b \left[\frac{\partial L}{\partial x} - \frac{d}{dt}\left(\frac{\partial L}{\partial x'} \right) \right] h(t)\, dt + \frac{\partial L}{\partial x'}\, h(t) \Big|_a^b.$$

For x to be an extremum we must have $\delta J(x,h) = 0$, identical with respect to $h(t)$. Since among the variations h are also those for which $h(a) = h(b) = 0$, using the remarks at the fundamental lemma, we obtain that the function $x(t)$ which performs the extremum must check the ODE (at most of second order in the unknown $x(t)$)

$$\frac{\partial L}{\partial x} - \frac{d}{dt}\left(\frac{\partial L}{\partial x'} \right) = 0. \tag{8}$$

This evolution ODE is called Euler–Lagrange ODE; its solutions $x(t)$ are called extremals (critical points) of the given functional.

So the functions $x(t)$ that perform an extremum of the functional are found among the extremals of the functional.

The Euler–Lagrange equation (8) is written explicitly

$$\frac{\partial^2 L}{\partial x' \partial x'}\, x'' + \frac{\partial^2 L}{\partial x \partial x'}\, x' + \frac{\partial^2 L}{\partial t \partial x'} - \frac{\partial L}{\partial x} = 0.$$

If this ODE with the unknown $x(t)$ is of second order, then the general solution depends on two arbitrary constants, so as to choose a certain solution we must impose conditions at the boundary.

Fixed ends extremals. *We impose to extremals given values at the ends of the integration interval: $x(a) = x_1$, $x(b) = x_2$. In this case we only have $h(a) = h(b) = 0$ and the necessary condition of extremum is reduced to Euler–Lagrange ODE (8). This, together with the conditions at the ends, generally determines a unique extremal.*

Extremals with free ordinates at the ends. *Let us suppose that only the point $x(a) = x_1$ is fixed. Then, the Euler–Lagrange ODE imposes the condition $\frac{\partial L}{\partial x'} h(b) = 0$. Since $h(b)$ is arbitrary, it is obtained a second condition $\frac{\partial L}{\partial x'}(b, x(b), x'(b)) = 0$. In this way we obtain a unique extremal. This equality is called natural boundary condition.*

Obviously, if we leave the ordinates free at both ends, we will get two natural boundary conditions.

Extremals with ends on given curves. *Let us suppose again $x(a) = x_1$, and the right end of the extremal be variable on a given curve $x = g(t)$. In this case the functional $J(x(\cdot))$ becomes an integral with a variable upper limit. For the functional variation we obtain, applying the formula for deriving an integral with a parameter,*

$$\delta J(x, h) = \frac{dJ}{d\varepsilon}(x + \varepsilon h)\Big|_{\varepsilon=0}$$

$$= \frac{d}{d\varepsilon} \int_a^{b(\varepsilon)} L(t, x(t) + \varepsilon h(t), x'(t) + \varepsilon h'(t)) dt \Big|_{\varepsilon=0}$$

$$= \int_a^b \left[\frac{\partial L}{\partial x}(t, x(t), x'(t))\, h(t) + \frac{\partial L}{\partial x'}(t, x(t), x'(t))\, h'(t) \right] dt$$

$$+ L(b, x(b), x'(b)) b'(\varepsilon)\Big|_{\varepsilon=0},$$

where b is given by the equality $x(b) = g(b)$.

To evaluate $b'(\varepsilon)|_{\varepsilon=0}$, we use the condition on the right end, that is

$$x(b(\varepsilon)) + \varepsilon h(b(\varepsilon)) = g(b(\varepsilon)).$$

Taking the derivative with respect to ε and setting $\varepsilon = 0$, we obtain

$$x'(b)\, b'(0) + h(b) = g'(b)\, b'(0),$$

whence

$$b'(0) = \frac{h(b)}{g'(b) - x'(b)}.$$

Then, after an integral by parts and selecting $h(a) = 0$, we find

$$\delta J(x, h) = \int_a^b \left[\frac{\partial L}{\partial x} - \frac{d}{dt}\left(\frac{\partial L}{\partial x'} \right) \right] h(t)\, dt$$

$$+ \left[\frac{\partial L}{\partial x'}(b, x(b), x'(b)) - \frac{L(b, x(b), x'(b))}{g'(b) - x'(b)} \right] h(b).$$

Since the variation h is arbitrary, we obtain, the Euler–Lagrange ODE (8) and then the transversality condition

$$\frac{\partial L}{\partial x'}(b, x(b), x'(b)) - \frac{L(b, x(b), x'(b))}{g'(b) - x'(b)} = 0 \tag{9}$$

(The transversality condition imposes $g'(b) - x'(b) \neq 0$, i.e., the extremal and the curve $x = g(t)$ are transversal at contact point.)

Obviously, if the left end of the extremal graph is variable on a given curve $x = f(t)$, we will have at this end as well a similar condition of transversality

$$\frac{\partial L}{\partial x'}(a, x(a), x'(a)) - \frac{L(a, x(a), x'(a))}{f'(a) - x'(a)} = 0,$$

where a is given by the equality $x(a) = f(a)$.

The results extend to functional with higher-order derivatives,

$$J(x(\cdot)) = \int_a^b L(t, x(t), x^{(1)}(t), ..., x^{(m)}(t))\, dt,$$

starting from the expression (3) of variation. Integrating by parts several times and using the fundamental lemma, the Euler–Lagrange equation takes the shape

$$\frac{\partial L}{\partial x} - \frac{d}{dt}\left(\frac{\partial L}{\partial x'}\right) + \frac{d^2}{dt^2}\left(\frac{\partial L}{\partial x''}\right) - ... + (-1)^m \frac{d^m}{dt^m}\left(\frac{\partial L}{\partial x^{(m)}}\right) = 0, \tag{10}$$

an ODE of order at most 2m, with unknown x(t).

Example 1.10: *For a functional that depend on several function variables*

$$J(x^1(\cdot), ..., x^n(\cdot)) = \int_a^b L(t, x^1(t), ..., x^n(t), \dot{x}^1(t), ..., \dot{x}^n(t))\, dt,$$

the extremals $x(t) = (x^1(t), ..., x^n(t))$ are solutions of Euler–Lagrange system

$$\frac{\partial L}{\partial x^i} - \frac{d}{dt}\left(\frac{\partial L}{\partial \dot{x}^i}\right) = 0, \quad i = \overline{1, n}. \tag{11}$$

The problems to which the extremals are constrained at the ends of the interval are posed and solved as in the previous case.

Example 1.11: *In case of double (multiple) integral functional*

$$J(w(\cdot)) = \iint_\Omega L(x, y, w(x, y), w_x(x, y), w_y(x, y))\, dx\, dy,$$

the variation is given by the relation (5). Integrating by parts and using Green formula we obtain

$$\delta J(w, h) = \iint_\Omega \left[\frac{\partial L}{\partial w} - \frac{\partial}{\partial x}\left(\frac{\partial L}{\partial w_x}\right) - \frac{\partial}{\partial y}\left(\frac{\partial L}{\partial w_y}\right)\right] h(x, y)\, dx\, dy$$

$$+ \iint_\Omega \left[\frac{\partial}{\partial x}\left(\frac{\partial L}{\partial w_x} h\right) + \frac{\partial}{\partial y}\left(\frac{\partial L}{\partial w_y} h\right)\right] dx\, dy$$

$$= \iint_\Omega \left[\frac{\partial L}{\partial w} - \frac{\partial}{\partial x}\left(\frac{\partial L}{\partial w_x}\right) - \frac{\partial}{\partial y}\left(\frac{\partial L}{\partial w_y}\right)\right] h(x, y)\, dx\, dy$$

$$+ \int_{\partial\Omega} h(x, y)\left(-\frac{\partial L}{\partial w_y}\, dx + \frac{\partial L}{\partial w_x}\, dy\right),$$

where $\partial\Omega$ represents the boundary of Ω.

If the function w is an extremum point, we must have $\delta J(w, h) = 0$, identically with respect to h. Considering that we have null variations h on the boundary $\partial\Omega$, we apply the *fundamental lemma as well as the accompanying remarks. We find that the function $w(x, y)$ must verify the Euler–Lagrange (Euler-Ostrogradski) PDE*

$$\frac{\partial L}{\partial w} - \frac{\partial}{\partial x}\left(\frac{\partial L}{\partial w_x}\right) - \frac{\partial}{\partial y}\left(\frac{\partial L}{\partial w_y}\right) = 0. \tag{12}$$

The solutions of this PDE are called extremal surfaces; any extremum point is an extremal surface.

Example 1.12: *Now let examine the extremals of curvilinear integral functional*

$$J(w(\cdot)) = \int_\Gamma [L_1(x, y, w(x, y), w_x(x, y), w_y(x, y))\, dx$$
$$+ L_2(x, y, w(x, y), w_x(x, y), w_y(x, y))\, dy\,],$$

where Γ is a piecewise C^1 curve which joins two fixed points A, B in a compact domain $\Omega \subset \mathbb{R}^2$. The variation of this functional is given by formula (6) which, after integration by parts, becomes

$$\delta J(w, h) = \int_\Gamma \left[\frac{\partial L_1}{\partial w} - \frac{\partial}{\partial x}\left(\frac{\partial L_1}{\partial w_x}\right) - \frac{\partial}{\partial y}\left(\frac{\partial L_1}{\partial w_y}\right)\right] h\, dx$$

$$+ \left[\frac{\partial L_2}{\partial w} - \frac{\partial}{\partial x}\left(\frac{\partial L_2}{\partial w_x}\right) - \frac{\partial}{\partial y}\left(\frac{\partial L_2}{\partial w_y}\right)\right] h\, dy$$

$$+ \int_\Gamma \left[\frac{\partial}{\partial x}\left(\frac{\partial L_1}{\partial w_x} h\right) + \frac{\partial}{\partial y}\left(\frac{\partial L_1}{\partial w_y} h\right)\right] dx + \left[\frac{\partial}{\partial x}\left(\frac{\partial L_2}{\partial w_x} h\right) + \frac{\partial}{\partial y}\left(\frac{\partial L_2}{\partial w_y} h\right)\right] dy.$$

To have an extremum it is necessary to cancel the variation $\delta J(w, h)$ identically with respect to h. For this purpose we use variations h, with $h(A) = h(B) = 0$, for which

$$\frac{\partial}{\partial y}\left(\frac{\partial L_1}{\partial w_y} h\right) = \frac{\partial}{\partial x}\left(\frac{\partial L_2}{\partial w_x} h\right).$$

The above system always has a solution being of type integral factor. For larger than 2 dimensions, the situation is a bit more complicated and will be solved in the following chapters.

The condition of critical point becomes

$$\int_\Gamma \left[\frac{\partial L_1}{\partial w} - \frac{\partial}{\partial x}\left(\frac{\partial L_1}{\partial w_x}\right) - \frac{\partial}{\partial y}\left(\frac{\partial L_1}{\partial w_y}\right)\right] h\, dx$$

$$+ \left[\frac{\partial L_2}{\partial w} - \frac{\partial}{\partial x}\left(\frac{\partial L_2}{\partial w_x}\right) - \frac{\partial}{\partial y}\left(\frac{\partial L_2}{\partial w_y}\right)\right] h\, dy = 0,$$

for any variation h in the previous conditions.

We will have to distinguish two situations as integral is taken on a fixed curve or the integral is path-independent. We will make, for abbreviation, the following notations:

$$EL_1 = \frac{\partial L_1}{\partial w} - \frac{\partial}{\partial x}\left(\frac{\partial L_1}{\partial w_x}\right) - \frac{\partial}{\partial y}\left(\frac{\partial L_1}{\partial w_y}\right),$$

$$EL_2 = \frac{\partial L_2}{\partial w} - \frac{\partial}{\partial x}\left(\frac{\partial L_2}{\partial w_x}\right) - \frac{\partial}{\partial y}\left(\frac{\partial L_2}{\partial w_y}\right),$$

$$EL = (EL_1, EL_2).$$

a) If the curvilinear integral is taken on a fixed path Γ : $x = x(t)$, $y = y(t)$, and $\tau = (\dot{x}, \dot{y})$ is the velocity field of the curve, the condition of critical point is the orthogonality on Γ of the vector field EL and τ, i.e.,

$$EL_1 \dot{x} + EL_2 \dot{y} = 0. \tag{13}$$

b) If the curvilinear functional is path-independent, the condition of critical point (13) must be satisfied for any vector $\tau = (\dot{x}, \dot{y})$ and then we obtain the PDEs system

$$\frac{\partial L_1}{\partial w} - \frac{\partial}{\partial x}\left(\frac{\partial L_1}{\partial w_x}\right) - \frac{\partial}{\partial y}\left(\frac{\partial L_1}{\partial w_y}\right) = 0,$$

$$\frac{\partial L_2}{\partial w} - \frac{\partial}{\partial x}\left(\frac{\partial L_2}{\partial w_x}\right) - \frac{\partial}{\partial y}\left(\frac{\partial L_2}{\partial w_y}\right) = 0, \tag{14}$$

with the unknown function $w(x, y)$.

1.3 Second Variation; Sufficient Conditions for Extremum

The third step in solving variational problems is the choice between the critical points of those who actually achieve an extremum. In concrete situations, coming from geometry, physics, engineering or economics, the realization of extrema is imposed by the nature of the problem.

Theoretically speaking, we need to find sufficient conditions for extremum. To this end we define the second variation of a functional.

Let us go back to the definition of differentiability, Definition 1.1, and suppose that the function f is derivable over its entire domain of definition D. Then the derivative operator $\mathbf{d}f(x, \cdot)$ becomes a function between two normed spaces

$$\mathbf{d}f(\cdot, \cdot) : D \subset \mathbf{U} \to \mathcal{L}(\mathbf{U}, \mathbf{V}), \quad x \mapsto \mathbf{d}f(x, \cdot),$$

where $\mathcal{L}(\mathbf{U}, \mathbf{V})$ is the normed space of linear bounded operators from $\mathbf{U}$ to $\mathbf{V}$. If this function is differentiable at a point $x \in D$, the derivative at x of the function $\mathbf{d}f$ is called *the second derivative of the function f at the point x*. The second derivative is a bilinear and continuous operator from $\mathbf{U} \times \mathbf{U}$ to $\mathbf{V}$ or, by identifying variations, a quadratic operator. Specifically, in the case of the functionals we dealt with previously, we have the following results.

Definition 1.2: If the variation $\delta J(x, h)$ is differentiable as function of x, its variation $\delta(\delta J(x, h))(k)$ is called the second variation of the functional J and is denoted by $\delta^2 J(x; h, k)$.

Proposition 1.3: *If the functional J is twice differentiable, then*

$$\delta^2 J(x; h, h) = \frac{d^2}{d\varepsilon^2} J(x + \varepsilon h)\Big|_{\varepsilon=0}.$$

On the other hand suppose that the deviation from the linear approximation, $r(x, h)$, small relative to h, can be approximated with a square operator, i.e.,

$$r(x, h) = A(h, h) + r_1(x, h), \ \lim_{h \to 0} \frac{r_1(x, h)}{\|h\|^2} = 0,$$

where A is a symmetric bilinear operator. Then the following property holds:
Second-order Taylor formula We have the equality

$$A(h, h) = \frac{1}{2} \delta^2 J(x; h, h)$$

and consequently

$$\triangle J(x(\cdot)) = J(x + h) - J(x) = \delta J(x, h) + \frac{1}{2} \delta^2 J(x; h, h) + r_1(x, h).$$

Next we will abbreviate $\delta^2 J(x; h, h)$ by $\delta^2 J(x, h)$
If x is a critical point (an extremal), i.e., $\delta J(x, h) = 0$, the nature of the critical point is given by the quadratic function $\delta^2 J(x, h)$. As in the case of real functions of real variables, it holds the following

Theorem 1.1: a) Necessary condition *If the extremal x is a minimum point of the functional J, then $\delta^2 J(x, h) \geq 0$ for any h.*
b) Sufficient condition *If for the extremal x the condition*

$$\delta^2 J(x, h) > C\|h\|^2 \ (C > 0 \ fixed)$$

holds for any h, then x is the minimum point for the functional J.

Remark 1.3: Condition b) cannot be weakened by $\delta^2 J(x, h) > 0$. For instance, if we consider the functional

$$J(x(\cdot)) = \int_0^1 x^2(t)(t - x(t)) \, dt$$

on the space $C^0[0, 1]$, then the null function $x(t) = 0, \forall t \in [0, 1]$, is a critical point and

$$\delta^2 J(x = 0, h) = \int_0^1 2t \, h^2(t) \, dt > 0$$

for any variation h. However J also has negative values in any neighborhood of the null function: for $\varepsilon > 0$ and the function $x_\varepsilon(t) = \varepsilon - t$, if $t < \varepsilon$, and null elsewhere,

$$J(x_\varepsilon(\cdot)) = \int_0^\varepsilon (\varepsilon - t)^2 (2t - \varepsilon)\, dt = -\frac{1}{6}\varepsilon^4.$$

Two examples of using the second-order variation.

Example 1.13: *For the functional $J(x(\cdot)) = \displaystyle\int_a^b L(t, x(t))\, dt$, the second variation is*

$$\delta^2 J(x, h) = \int_a^b \frac{\partial^2 L}{\partial x^2}(t, x(t))\, h^2(t)\, dt.$$

On the other hand, developing with respect to the powers of h, around the critical point x, we find

$$\Delta J(x, h) = \int_a^b \left(\frac{1}{2}\frac{\partial^2 L}{\partial x^2}(t, x(t))\, h^2(t) + r_2(x, h)\, h^3(t) \right) dt$$

$$= \int_a^b h^2(t) \left(\frac{1}{2}\frac{\partial^2 L}{\partial x^2}(t, x(t)) + r_2(x, h)\, h(t) \right) dt,$$

where $r_2(x, \cdot)$ is a bounded function around $h = 0$. It follows that the critical point x is the minimum point of the functional J, if and only if

$$\frac{\partial^2 L}{\partial x^2}(t, x(t)) > 0, \quad \forall t \in [a, b].$$

Example 1.14: *For the functional $J(x(\cdot)) = \displaystyle\int_a^b L(t, x(t), x'(t))\, dt$, the second variation is*

$$\delta^2 J(x, h) = \frac{1}{2}\int_a^b \left(\frac{\partial^2 L}{\partial x^2}h^2 + 2\frac{\partial^2 L}{\partial x \partial x'}hh' + \frac{\partial^2 L}{\partial x'^2}h'^2 \right) dt.$$

Considering that $2hh' = (h^2)'$ and integrating the second term by parts, the second variation becomes

$$\delta^2 J(x, h) = \int_a^b \left(F(t)\, h^2 + \frac{1}{2}\frac{\partial^2 L}{\partial x'^2}h'^2 \right) dt, \tag{15}$$

where we noted $F(t) = \dfrac{1}{2}\left(\dfrac{\partial^2 L}{\partial x^2} - \dfrac{d}{dx}\left(\dfrac{\partial^2 L}{\partial x \partial x'} \right) \right)(t).$

Theorem 1.2: **(Legendre's necessary condition)** *If the extremal function x is the minimum point for J, then*

$$\frac{\partial^2 L}{\partial x'^2}(t, x(t), x'(t)) \geq 0, \quad \forall t \in [a, b].$$

Proof. We use the same procedure as in the case of the fundamental lemma. We assume that at a point $t_0 \in [a, b]$ we have

$$\frac{\partial^2 L}{\partial x'^2}(t_0, x(t_0), x'(t_0)) < 0.$$

Then the same inequality will take place in a neighborhood U of the point t_0. We construct a C^1 variation $h(t_0 + \tau)$, which has values between 0 and 1, that cancels outside the neighborhood U and $h(t_0) = 1$. In addition, we will require that the derivative of the function h to be large enough in module at certain intervals. For example, if $U = (t_0 - \varepsilon, t_0 + \varepsilon)$, we can take $h(t_0 + \tau) = \frac{1}{\varepsilon^4}(m\tau + \varepsilon)^2(\varepsilon - m\tau)^2$, for $|\tau| < \frac{\varepsilon}{m}$ and 0 elsewhere. A direct calculation gives us

$$\int_{|\tau| < \frac{\varepsilon}{m}} h(t_0 + \tau) \, d\tau = \frac{256}{105} \frac{m}{\varepsilon}.$$

If $\frac{\partial^2 L}{\partial x'^2}(t, x(t), x'(t)) \leq -C < 0$, in U, then we have

$$\delta^2 J(x, h) \leq \int_a^b |F(t)| \, dt - \frac{256}{105} \frac{m}{\varepsilon} C$$

and then we can choose m so that $\delta^2 J(x, h) < 0$, which contradicts the necessary minimum condition.

Legendre's condition extends to functions of several variables. Thus, *if the extremum function w is the minimum point for the double integral functional*

$$J(w(\cdot)) = \iint_\Omega L(x, y, w(x, y), w_x(x, y), w_y(x, y)) \, dx \, dy,$$

then the quadratic form

$$Q(h) = w_{xx}(x, y) (h^1)^2 + 2w_{xy}(x, y) h^1 h^2 + w_{yy}(x, y)(h^2)^2$$

is positive semidefinite at any point in Ω (the notation $w_{..}$ represents second-order derivatives).

Specific and comfortable sufficient conditions for extremum are difficult to obtain; some, more or less classic, will be given in the following chapters.

1.4 Optimum with Constraints; the Principle of Reciprocity

In the previous problems the unknown functions were arbitrary in the set of definition of the functional, being subject only to certain border conditions. We will continue to deal with issues in which unknown functions are subject to conditions regarding their entire behavior.[1]

---

1 Chronologically, it seems that the first problem of this type was Dido's problem (ninth century BC): Among all the closed curves of given length to find the one that closes a maximum area.

1.4.1 Isoperimetric Problems

The simplest isoperimetric problem is

Problem 1.4.1: Find the function $x = x(t)$, extremum for the functional

$$J(x(\cdot)) = \int_a^b L(t, x(t), x'(t)) \, dt, \tag{16}$$

among the functions which have given values at the ends of the interval $x(a) = x_1, x(b) = x_2$ and check the additional condition

$$K(x) = \int_a^b G(t, x(t), x'(t)) \, dt = c \ (constant). \tag{17}$$

Solution. Suppose that $x = x(t)$ is the function sought and consider the variation $\varepsilon h + \eta k$ such that $K(x + \varepsilon h + \eta k) = c$, where ε and η are parameters (the family of variations depends on two parameters, because the variation with a single parameter does not generally satisfy the condition $K = constant$), and h and k are functions for which $h(a) = h(b) = k(a) = k(b) = 0$.

The isoperimetric problem now returns to the statement: The function of two real variables

$$J(\varepsilon, \eta) = \int_a^b L(t, x(t) + \varepsilon h(t) + \eta k(t), x'(t) + \varepsilon h'(t) + \eta k'(t)) \, dt$$

has at the point $(0, 0)$ an extremum conditioned by the restriction

$$K(\varepsilon, \eta) = \int_a^b G(t, x(t) + \varepsilon h(t) + \eta k(t), x'(t) + \varepsilon h'(t) + \eta k'(t)) \, dt = c.$$

Lagrange's multiplier theory states that then there is a number λ such that the point $(0, 0)$ is a critical point for the function $\Phi(\varepsilon, \eta) = J(\varepsilon, \eta) + \lambda K(\varepsilon, \eta)$. Critical point conditions

$$\frac{\partial \Phi}{\partial \varepsilon}(0, 0) = 0, \ \frac{\partial \Phi}{\partial \eta}(0, 0) = 0,$$

give us, *via* integrations by parts, respectively

$$\int_a^b \left[\left(\frac{\partial L}{\partial x} - \frac{d}{dt}\left(\frac{\partial L}{\partial x'}\right) \right) + \lambda \left(\frac{\partial G}{\partial x} - \frac{d}{dt}\left(\frac{\partial G}{\partial x'}\right) \right) \right]_{(t,x(t),x'(t))} h(t) \, dt = 0,$$

$$\int_a^b \left[\left(\frac{\partial L}{\partial x} - \frac{d}{dt}\left(\frac{\partial L}{\partial x'}\right) \right) + \lambda \left(\frac{\partial G}{\partial x} - \frac{d}{dt}\left(\frac{\partial G}{\partial x'}\right) \right) \right]_{(t,x(t),x'(t))} k(t) \, dt = 0,$$

for any variations h and k, null at a and b.

Consequently the right parenthesis in the above integrals must be canceled and we get the next one

Theorem 1.3: *If the real number c is not an extreme value for the integral (17), then the solutions of the isoperimetric problem (16) + (17) are extremal functions, without additional conditions, for the functional*

$$\mathcal{J}(x(\cdot)) = \int_a^b \mathcal{L}(t, x(t), x'(t)) \, dt,$$

where $\mathcal{L}(t, x(t), x'(t)) = L(t, x(t), x'(t)) + \lambda G(t, x(t), x'(t))$.

Proof. Suppose that the Euler–Lagrange ODE

$$\frac{\partial \mathcal{L}}{\partial x} - \frac{d}{dt}\left(\frac{\partial \mathcal{L}}{\partial x'}\right) = 0 \tag{18}$$

is of second order (at most) in the unknown function $x(t)$. Integrating, we obtain the general solution depending on λ and two more arbitrary constants. The three constants are determined from the restriction (17) and the conditions at the ends of the interval. □

1.4.2 The Reciprocity Principle

Equation (18) also solves the following problem: *among all the curves joining two given points, determine the curve at the variation of which one has* $\delta J = 0$ *as soon as* $\delta K = 0$ (the curves which solve such a problem are called *conditioned extremals*).

Indeed, if the function $x(\cdot)$ verifies equation (18), we have $\delta(J + \lambda K) = 0$; but then the condition $\delta K = 0$ implies $\delta J = 0$, that is x is a conditioned extremal.

Conversely, a simple property of the general theory of linear functionals is used: if the linear functionals f and g vanish at the same vectors, then the two functionals are proportional:

$$f(h) = \lambda g(h), \quad \forall h \quad (\lambda \ \ fixed).$$

Returning to the isoperimetric problem, let us observe that by multiplying the function L in the integral (16) by a nonzero constant, the family of extremals of the integral does not change; then we can write the Lagrangian $\mathcal{L}$ in a symmetrical form

$$\mathcal{L} = \lambda_1 L + \lambda_2 G,$$

where λ_1 and λ_2 are constants. In this representation the functions L and G have symmetrical roles. If we exclude the situation $\lambda_1 = 0$ or $\lambda_2 = 0$, we obtain **the reciprocity principle**: *problem* J = *extremum with the condition* K = *constant has the same family of extremals as the problem* K = *extremum with the condition* J = *constant*.

1.4.3 Constrained Extrema: The Lagrange Problem

Lagrange's multiplier method is also applicable in cases where we take as admissible curves, curves located on a surface, integral curves of a system of differential equations or curves located in sets of constant level of some functional ones. This situation is called *Lagrange's (general) problem*.

One such problem is, for example: Finding the extremum of the functional

$$J(x(\cdot), y(\cdot)) = \int_a^b L(t, x(t), y(t), x'(t), y'(t)) \, dt, \tag{19}$$

in the set of class C^1 curves, checking the differential relation

$$G(t, x(t), y(t), x'(t), y'(t)) = 0 \tag{20}$$

and with certain additional conditions at the extremities. To solve this problem, we use the following:

Theorem 1.4: *If the curve $\gamma_0 = (x, y)$ realizes the extremum of the functional (19) with the condition (20) and if, on γ_0, at least one of the derivatives $G_{x'}$ or $G_{y'}$ is not canceled, then there is a function $\lambda(t)$, so that γ_0 is extremum, without additional conditions for functional*

$$\mathcal{J}(x(\cdot), y(\cdot)) = \int_a^b \mathcal{L}(t, x(t), y(t), x'(t), y'(t)) \, dt,$$

where $\mathcal{L}(t, x, y, x', y') = L(t, x, y, x', y') + \lambda(t)G(t, x, y, x', y')$.

Proof. Indeed, by solving the differential system, assumed of the second order,

$$\frac{\partial \mathcal{L}}{\partial x} - \frac{d}{dt}\left(\frac{\partial \mathcal{L}}{\partial x'}\right) = 0, \quad \frac{\partial \mathcal{L}}{\partial y} - \frac{d}{dt}\left(\frac{\partial \mathcal{L}}{\partial y'}\right) = 0, \tag{21}$$

to which is added equation (20), we find the unknown functions x, y, λ. These depend on four arbitrary constants, because the system naturally transforms into a system of four first-order equations in the variables λ, x, y, u, v between which there is the relation (20): $G(t, x, y, u, v) = 0$. The conditions at the ends of interval determine these four constants.

Explanatory note: Even if from the ODE (20) we can replace x' and y' in (19), the constraint optimum character is preserved. □

Although the theorem has a specific proof, which we do not give, we notice that there is a result of the functional analysis, similar with the one mentioned in the reciprocity principle: Let **X** and **Y** be two Banach spaces, $J : \mathbf{X} \to \mathbf{R}$ a differentiable functional and $G : \mathbf{X} \to \mathbf{Y}$ a differentiable operator. If $x_0 \in \mathbf{X}$ is a point where *(i)* the variation $\delta G(x_0, \cdot) : \mathbf{X} \to \mathbf{Y}$ is surjective, *(ii)* $\delta J(x_0, h) = 0$ as soon as $\delta G(x_0, h) = 0$, then there exists a linear and continuous functional $\lambda : \mathbf{Y} \to \mathbf{R}$ such that

$$\delta J(x_0, \cdot) = \lambda \circ \delta G(x_0, \cdot). \tag{22}$$

In the case of relation (20), **Y** is the space $C^0[a, b]$ of the continuous functions and then

$$\lambda \circ \delta G(x_0, y_0, h) = \int_a^b \lambda(t) \, \delta G(x_0, y_0, h) \, dt.$$

1.5   Maple Application Topics

Maple commands

The most useful Maple command for this section will be the solve command. In order to find the critical points of the function $f(x, y) = y^2 - x^2$, you would first find the partial derivatives fx and fy using the Maple diff command. Then, you solve $fx = 0$ and $fy = 0$ using Maple. The following sequence will find critical points of $f(x, y)$ defined above.

> $f := y^2 - x^2$;
> $fx := diff(f, x)$;
> $fy := diff(f, y)$;
> cp:=solve($\{fx = 0, fy = 0\}, \{x, y\}$);

The second derivative test can be applied by differentiating f twice with respect to x and y, and finding fxy as well. The following Maple commands find the discriminant function d and evaluate the discriminant function at each of the critical points. The second command eval(fxx, cp) evaluates fxx at the critical point so that you can determine if the critical point is a local minimum, a local maximum or a saddle point.

> $fxx := diff(f, x, x)$;
> $d := diff(f, x, x) * diff(f, y, y) - diff(f, x, y)^2$;
> eval(d, cp); eval(fxx, cp);

Problem 1.5.1:   > extrema($axyz, x^2 + y^2 + z^2 = 1, \{x, y, z\}$);

$$\left\{ \max\left(0, -\frac{\sqrt{3}}{9}\,a, \frac{\sqrt{3}}{9}\,a\right), \min\left(0, -\frac{\sqrt{3}}{9}\,a, \frac{\sqrt{3}}{9}\,a\right) \right\}.$$

Problem 1.5.2:   > ExtremePoints($x^2 - 3x + x, x$); $[-1, 1]$

Calling Sequence

VariationalCalculus[command](arguments)

The VariationalCalculus package contains commands that perform calculus of variations computations. This package performs various computations for characterizing the extremals of functionals.

EulerLagrange

Problem 1.5.3:   Find the geodesics ODE in the plane:

> with(VariationalCalculus);
$L := \sqrt{(diff(x(t), t)^2 + diff(y(t), t)^2}$;
> EulerLagrange($L, t, [x(t), y(t)]$);

Problem 1.5.4:   Create a space of three independent variables and one dependent variable. Derive the Laplace's equation from its variational principle.

Solution.   > $with(DifferentialGeometry); with(JetCalculus)$;

E > DGsetup($[x, y, z], [u], E, 1$):

E > $L = \frac{1}{2}(u_1^2 + u_2^2 + u_3^2)$

E > E_3:= EulerLagrange(L)

E > convert($E3[1], DGdiff$)

Answer: $-u_{x,x} - u_{y,y} - u_{z,z} = 0$.

Repeat this computation using differential forms.

E > $\lambda 3 := evalDG(L *`\&w`(`\&w`(Dx, Dy), Dz))$;

E > EulerLagrange($lambda3$)

Problem 1.5.5: Create a space of three independent variables and three dependent variables. Derive three-dimensional Maxwell equations from the variational principle.

Solution. Define the Lagrangian M > $L := -(1/2) * A_t[2]^2 + A_t[2] * A_y[3] - (1/2) * A_y[3]^2 - (1/2) * A_t[1]^2 + A_t[1] * A_x[3] - (1/2) * A_x[3]^2 + (1/2) * A_y[1]^2 - A_y[1] * A_x[2] + (1/2) * A_x[2]^2$

Compute the Euler–Lagrange equations.

M > Maxwell1 := EulerLagrange(L);

Change notation to improve readability

M > PDEtools[declare](quiet);

M > Maxwell2 := map(convert, Maxwell1, DGdiff);

2

Variational Principles

Motto:
"When my angel nail was blunted
I let it grow again,
But it didn't,
Or else I knew nothing of it."
       Tudor Arghezi – *Flowers of Mildew*

In this chapter, we will study adapted variational problems, together with boundary conditions, subsidiary conditions or restrictions imposed on the class of admissible solutions that produce extrema values for a given functional. We will also insist on sufficient conditions (Legendre-Jacobi test and invexity test) for the existence of the minimum or maximum of a functional. Details about the variational principles can be found in the papers [4, 6, 7, 12, 15, 16, 18, 19, 21, 22, 30, 31, 33, 34, 48, 57, 58, 71, 79, 80, 83].

2.1 Problems with Natural Conditions at the Boundary

Problem 2.1.1: Let $L(x, y, y')$ be a C^2 Lagrangian. Find the C^2 function $x \mapsto y(x)$ for which the functional

$$I(y(\cdot)) = \int_{x_1}^{x_2} L(x, y(x), y'(x)) \, dx$$

has an extremum value, when one or both boundary conditions $y(x_1) = y_1, y(x_2) = y_2$ are not prescribed.

Solution. Suppose the function $y = y(x)$ produces an extremum value and we build a variation $y_\varepsilon(x) = y(x) + \varepsilon h(x)$ of class C^2. We substitute in the functional and we obtain an integral with a parameter (function)

$$I(\varepsilon) = \int_{x_1}^{x_2} L(x, y(x) + \varepsilon h(x), y'(x) + \varepsilon h'(x)) \, dx.$$

By hypothesis $\varepsilon = 0$ is an extremum value of $I(\varepsilon)$. So, applying the derivation under the integral sign, we find

Variational Calculus with Engineering Applications, First Edition. Constantin Udriste and Ionel Tevy.
© 2023 John Wiley & Sons Ltd. Published 2023 by John Wiley & Sons Ltd.

$$0 = I'(0) = \int_{x_1}^{x_2} \left(\frac{\partial L}{\partial y} h + \frac{\partial L}{\partial y'} h' \right) dx.$$

Integrating by parts the second term in the integral, we deduce

$$0 = I'(0) = \frac{\partial L}{\partial y'} h \Big|_{x_1}^{x_2} + \int_{x_1}^{x_2} \left(\frac{\partial L}{\partial y} - \frac{d}{dx} \frac{\partial L}{\partial y'} \right) h \, dx. \tag{1}$$

Suppose that $y = y(x)$ is the solution of Euler–Lagrange equation

$$\frac{\partial L}{\partial y} - \frac{d}{dx} \frac{\partial L}{\partial y'} = 0$$

(differential equation of at most second order) on the interval $[x_1, x_2]$. In order for relation (1) to be satisfied it is necessary that

$$\frac{\partial L}{\partial y'} h \Big|_{x_1}^{x_2} = \frac{\partial L}{\partial y'}(x_2) h(x_2) - \frac{\partial L}{\partial y'}(x_1) h(x_1) = 0.$$

In case of fixed points $y(x_1) = y_1$, $y(x_2) = y_2$, the Euler–Lagrange equation together with $h(x_1) = h(x_2) = 0$ guarantee the relationship (1). If both ends are variable, the values $h(x_1), h(x_2)$ are arbitrary and for the relation (1) to be true we must impose the Euler–Lagrange equation together with the conditions

$$\frac{\partial L}{\partial y'}(x_1) = 0, \quad \frac{\partial L}{\partial y'}(x_2) = 0. \tag{2}$$

These conditions are called *natural conditions at the boundary* or *transversability conditions* for the extremum problem. The following cases occur naturally:

(i) **fixed endpoints** If the ends of the graph $y(x_1) = y_1$, $y(x_2) = y_2$ are fixed, then there is no variation in these ends, i.e., $h(x_1) = h(x_2) = 0$.
(ii) **mixed boundary conditions** If $y(x_1) = y_1$ is fixed and $y(x_2) = y_2$ is arbitrary, then we must impose $h(x_1) = 0$, $\frac{\partial L}{\partial y'}(x_2) = 0$.
(iii) **mixed boundary conditions** If $y(x_1) = y_1$ is arbitrary and $y(x_2) = y_2$ is fixed, then we must impose $\frac{\partial L}{\partial y'}(x_1) = 0$, $h(x_2) = 0$.
(iv) **variable ends** If the points $y(x_1) = y_1$ and $y(x_2) = y_2$ are arbitrary, then we must impose conditions (2).

The boundary conditions in which the h variation is canceled are called *essential boundary conditions, or Dirichlet-type boundary conditions, or geometric boundary conditions*. Natural boundary conditions are also called *Neumann-type boundary conditions, or dynamic boundary conditions*. Obviously we can have mixed conditions at the border. The terminology here extends to multidimensional problems.

Problem 2.1.2: Let $L(x, y, y', y'')$ be a C^3 Lagrangian. Find the C^4 function $x \mapsto y(x)$ for which the functional

$$I(y(\cdot)) = \int_{x_1}^{x_2} L(x, y(x), y'(x), y''(x))\, dx$$

has an extremum value.

Solution. Let $\delta y = y_1 - y = h$ be a variation in y. If y is the solution of the problem, then the first variation of the functional is

$$\delta I = \int_{x_1}^{x_2} \left(\frac{\partial L}{\partial y} \delta y + \frac{\partial L}{\partial y'} \delta y' + \frac{\partial L}{\partial y''} \delta y'' \right) dx = 0.$$

We use the integration by parts. For the second term under integral, we set

$$u = \frac{\partial L}{\partial y'}, \quad du = \frac{d}{dx}\left(\frac{\partial L}{\partial y'} \right) dx, \quad dv = \delta y' dx, \quad v = \delta y,$$

and analogously for the third term. The variation δI is written in the form

$$\int_{x_1}^{x_2} \left(\frac{\partial L}{\partial y} \delta y - \frac{d}{dx}\left(\frac{\partial L}{\partial y'} \right) \delta y - \frac{d}{dx}\left(\frac{\partial L}{\partial y''} \right) \delta y' \right) dx$$

$$+ \frac{\partial L}{\partial y'} \delta y \Big|_{x_1}^{x_2} + \frac{\partial L}{\partial y''} \delta y' \Big|_{x_1}^{x_2} = 0.$$

Again we integrate by parts with respect to the third term under integral. We find

$$\delta I = \int_{x_1}^{x_2} \left(\frac{\partial L}{\partial y} - \frac{d}{dx}\left(\frac{\partial L}{\partial y'} \right) + \frac{d^2}{dx^2}\left(\frac{\partial L}{\partial y''} \right) \right) \delta y\, dx$$

$$+ \left(\frac{\partial L}{\partial y'} - \frac{d}{dx}\left(\frac{\partial L}{\partial y''} \right) \right) \delta y \Big|_{x_1}^{x_2} + \frac{\partial L}{\partial y''} \delta y' \Big|_{x_1}^{x_2} = 0.$$

The Euler–Lagrange ODE

$$\frac{\partial L}{\partial y} - \frac{d}{dx}\frac{\partial L}{\partial y'} + \frac{d^2}{dx^2}\frac{\partial L}{\partial y''} = 0$$

is a differential equation at most of order four in the unknown $y(x)$. The boundary conditions are

$$\left(\frac{\partial L}{\partial y'} - \frac{d}{dx}\left(\frac{\partial L}{\partial y''} \right) \right) \delta y \Big|_{x_1}^{x_2} = 0, \quad \frac{\partial L}{\partial y''} \delta y' \Big|_{x_1}^{x_2} = 0.$$

If $y(x_1) = y_1$, $y(x_2) = y_2$, $y'(x_1) = y_1^1$, $y'(x_2) = y_2^1$ are fixed, then δy and $\delta y'$ vanish at x_1 and x_2. Alternatively, we can have the natural boundary conditions

$$\left(\frac{\partial L}{\partial y'} - \frac{d}{dx}\left(\frac{\partial L}{\partial y''} \right) \right) \Big|_{x_1}^{x_2} = 0, \quad \frac{\partial L}{\partial y''} \Big|_{x_1}^{x_2} = 0.$$

Obviously, the combinations lead to mixed boundary conditions.

Example 2.1: *Find the function $x \mapsto y(x)$ of class C^4 for which the functional*

$$I(y(\cdot)) = \int_{x_1}^{x_2} \left(a(x)(y'')^2(x) - b(x)(y')^2(x) + c(x)y^2(x) \right) dx$$

has an extremum value.

Solution. The Lagrangian $L = a(x)(y'')^2 - b(x)(y')^2 + c(x)y^2$ has the partial derivatives

$$\frac{\partial L}{\partial y} = 2c(x)y, \quad \frac{\partial L}{\partial y'} = -2b(x)y', \quad \frac{\partial L}{\partial y''} = 2a(x)y''.$$

It follows Euler–Lagrange ODE (fourth-order)

$$c(x)y + \frac{d}{dx}(b(x)y') + \frac{d^2}{dx^2}(a(x)y'') = 0, \; x_1 \le x \le x_2.$$

To this are added the boundary conditions:

(i) if δy and $\delta y'$ vanish at the ends, then $y(x_1) = y_1$, $y(x_2) = y_2$, $y'(x_1) = y_1^1$, $y'(x_2) = y_2^1$ must be prescribed;

(ii) if δy vanishes at ends, and $\delta y'$ does not vanish at ends, then $y(x_1) = y_1$, $y(x_2) = y_2$ must be prescribed and the natural conditions $\frac{\partial L}{\partial y''}(x_1) = 0$, $\frac{\partial L}{\partial y''}(x_2) = 0$ must be verified; in this case $y''(x_1)$ and $y''(x_2)$ have fixed values;

(iii) if δy does not vanish at ends, and $\delta y'$ vanishes at ends, then $y'(x_1) = y_1^1$, $y'(x_2) = y_2^1$ must be prescribed and the natural conditions

$$\left(\frac{\partial L}{\partial y'} - \frac{d}{dx}\left(\frac{\partial L}{\partial y''} \right) \right)(x_1) = 0, \; \left(\frac{\partial L}{\partial y'} - \frac{d}{dx}\left(\frac{\partial L}{\partial y''} \right) \right)(x_2) = 0$$

must be satisfied;

(iv) if the variations δy and $\delta y'$ do not vanish at ends, then the boundary natural conditions are

$$\frac{\partial L}{\partial y''}(x_1) = 0, \; \frac{\partial L}{\partial y''}(x_2) = 0$$

$$\left(\frac{\partial L}{\partial y'} - \frac{d}{dx}\left(\frac{\partial L}{\partial y''} \right) \right)(x_1) = 0, \; \left(\frac{\partial L}{\partial y'} - \frac{d}{dx}\left(\frac{\partial L}{\partial y''} \right) \right)(x_2) = 0.$$

Problem 2.1.3: Let $L(x, y, w, w_x, w_y)$ be a C^2 Lagrangian. Let $\Omega \subset \mathbb{R}^2$ with simple closed boundary $\partial\Omega$ of piecewise C^1 class. Find the function $(x, y) \mapsto w(x, y)$ of class C^2 for which the double integral functional

$$I(w(\cdot)) = \int_{\Omega} L(x, y, w(x, y), w_x(x, y), w_y(x, y)) \, dx \, dy$$

has an extremum value.

Solution. We use the variational notations. We require that the first variation be zero,

$$\delta I = \int_\Omega \left(\frac{\partial L}{\partial w} \delta w + \frac{\partial L}{\partial w_x} \delta w_x + \frac{\partial L}{\partial w_y} \delta w_y \right) dx \, dy = 0.$$

For $N = \dfrac{\partial L}{\partial w_x} \delta w$ and $M = -\dfrac{\partial L}{\partial w_y} \delta w$, we apply the Green Theorem in plane,

$$\int_\Omega \left(\frac{\partial N}{\partial x} - \frac{\partial M}{\partial y} \right) dx \, dy = \int_{\partial \Omega} M \, dx + N \, dy.$$

Since

$$\frac{\partial N}{\partial x} = \frac{\partial L}{\partial w_x} \delta w_x + \frac{\partial}{\partial x} \left(\frac{\partial L}{\partial w_x} \right) \delta w, \quad \frac{\partial M}{\partial y} = -\frac{\partial L}{\partial w_y} \delta w_y - \frac{\partial}{\partial y} \left(\frac{\partial L}{\partial w_y} \right) \delta w,$$

we find

$$\delta I = \int_\Omega \left(\frac{\partial L}{\partial w} - D_x \frac{\partial L}{\partial w_x} - D_y \frac{\partial L}{\partial w_y} \right) \delta w \, dx \, dy$$

$$+ \int_{\partial \Omega} \left(-\frac{\partial L}{\partial w_y} dx + \frac{\partial L}{\partial w_x} dy \right) \delta w = 0.$$

In this way the function $w(x, y)$ is a solution of Euler–Lagrange PDE

$$\frac{\partial L}{\partial w} - D_x \frac{\partial L}{\partial w_x} - D_y \frac{\partial L}{\partial w_y} = 0$$

(at most of second order) to which we add boundary conditions. If $\delta w = 0$, the function $w(x, y)$ must be specified on $\partial \Omega$, i.e., $w(x, y)\big|_{\partial \Omega}$ = given (Dirichlet conditions). If $\delta w \neq 0$, then we impose the boundary natural condition (Newmann conditions)

$$-\frac{\partial L}{\partial w_y} dx + \frac{\partial L}{\partial w_x} dy \bigg|_{(x,y) \in \partial \Omega} = 0.$$

The last condition can be written using the unit normal vector at the boundary $\partial \Omega : \vec{\gamma} = x(s)\vec{i} + y(s)\vec{j}$. If $\frac{d\vec{\gamma}}{ds} = \vec{t} = \frac{dx}{ds}\vec{i} + \frac{dy}{ds}\vec{j}$ is the tangent versor, then $\vec{n} = \frac{dy}{ds}\vec{i} - \frac{dx}{ds}\vec{j}$ is the unit normal vector. With these, the natural condition at the boundary is written

$$\left\langle \vec{n}, \frac{\partial L}{\partial w_x}\vec{i} + \frac{\partial L}{\partial w_y}\vec{j} \right\rangle \bigg|_{(x,y) \in \partial \Omega} = 0.$$

2.2 Sufficiency by the Legendre-Jacobi Test

Traditionally, a sufficient condition for the existence of an extremum of a functional is considered to be the Legendre-Jacobi test. We further add the test of invexity, which appeared in the newer mathematical literature.

Let $L(x, y, y')$ be a C^2 Lagrangian associated to the C^2 function $x \mapsto y(x)$. We consider again the functional

$$I(y(\cdot)) = \int_{x_1}^{x_2} L(x, y(x), y'(x)) \, dx$$

and suppose that $y = y(x)$ is an extremum point that satisfies the conditions $y(x_1) = y_1$, $y(x_2) = y_2$. Let $y_\varepsilon(x) = y(x) + \varepsilon h(x)$ be a C^2 variation which satisfies $h(x_1) = h(x_2) = 0$. Then the functional $I(y(\cdot))$ is replaced by the integral $I(\varepsilon)$. We develop this in Taylor series using powers of ε (around the point $\varepsilon = 0$). We find

$$I(\varepsilon) = \int_{x_1}^{x_2} L(x, y_\varepsilon(x), y_\varepsilon'(x)) \, dx = I(0) + I'(0)\varepsilon + I''(0)\frac{\varepsilon^2}{2} + \dots$$

On the other hand, the first two derivatives $I'(0)$, $I''(0)$ are

$$I'(0) = \int_{x_1}^{x_2} \left(\frac{\partial L}{\partial y} h + \frac{\partial L}{\partial y'} h' \right) dx,$$

$$I''(0) = \int_{x_1}^{x_2} \left(\frac{\partial^2 L}{\partial y^2} h^2 + 2 \frac{\partial^2 L}{\partial y \partial y'} hh' + \frac{\partial^2 L}{\partial y'^2} h'^2 \right) dx.$$

We deal with the first-order variation. Integrating through parts, it is written

$$I'(0) = \frac{\partial L}{\partial y'} h \Big|_{x_1}^{x_2} + \int_{x_1}^{x_2} \left(\frac{\partial L}{\partial y} - \frac{d}{dx} \left(\frac{\partial L}{\partial y'} \right) \right) h \, dx.$$

Since h is an arbitrary variation with $h(x_1) = h(x_2) = 0$, from the condition $I'(0) = 0$ we find the necessary conditions

$$\frac{\partial L}{\partial y} - \frac{d}{dx} \left(\frac{\partial L}{\partial y'} \right) = 0, \; y(x_1) = y_1, \; y(x_2) = y_2.$$

If so, the sign of the difference $I(\varepsilon) - I(0)$ is given by $I''(0)$. If we have $I''(0) > 0$, on extremal arc, then $I(0)$ is a minimum; if $I''(0) < 0$, on extremal arc, then $I(0)$ is maximum; if $I''(\varepsilon)$ changes its sign around $\varepsilon = 0$, then $I(0)$ is not an extremum; if $I''(0) = 0$ on extremal arc, then we appeal to higher-order derivatives.

Let us analyze the second variation $I''(0)$, knowing $h(x_1) = h(x_2) = 0$ and $\frac{\partial L}{\partial y} - \frac{d}{dx} \left(\frac{\partial L}{\partial y'} \right) = 0$, $y(x_1) = y_1$, $y(x_2) = y_2$. If $y(x)$ is the solution of Euler–Lagrange ODE and if $I(y)$ is a minimum value, then $I''(0) \geq 0$; if $y(x)$ is the solution of Euler–Lagrange ODE and if $I(y)$ is maximum value, then $I''(0) \leq 0$.

The second variation $I''(0)$ can be transcribed in different forms.

First of all, integrating the middle term by parts, we find

$$I''(0) = h^2 \frac{\partial^2 L}{\partial y \partial y'}\Big|_{x_1}^{x_2} + \int_{x_1}^{x_2} \left(\left(\frac{\partial^2 L}{\partial y^2} - \frac{d}{dx} \frac{\partial^2 L}{\partial y \partial y'} \right) h^2 + \frac{\partial^2 L}{\partial y'^2} h'^2 \right) dx.$$

If we take $h(x_1) = h(x_2) = 0$, then

$$\frac{\partial^2 L}{\partial y^2} - \frac{d}{dx} \frac{\partial^2 L}{\partial y \partial y'} > 0, \quad \frac{\partial^2 L}{\partial y'^2} > 0$$

are necessary and sufficient conditions for minimum.

Secondly, we write

$$I''(0) = \int_{x_1}^{x_2} \left((L_{yy} h^2 + L_{yy'} hh') + h' (L_{yy'} h + L_{y'y'} h') \right) dx.$$

We integrate the second term by parts

$$I''(0) = \left(h^2 L_{yy'} + hh' L_{y'y'} \right)\Big|_{x_1}^{x_2}$$

$$+ \int_{x_1}^{x_2} \left(h^2 L_{yy} - h^2 \frac{d}{dx} L_{yy'} - h \frac{d}{dx} \left(h' L_{y'y'} \right) \right) dx.$$

The first term is null by hypotheses. Denoting

$$A(h) = \frac{d}{dx} \left(L_{y'y'} h' \right) - \left(L_{yy} - \frac{d}{dx} L_{yy'} \right) h,$$

the second term can be written in the form

$$I''(0) = - \int_{x_1}^{x_2} h A(h) \, dx.$$

The differential operator A has the form

$$A(u) = \frac{d}{dx} \left(p(x) \frac{du}{dx} \right) - q(x) u,$$

where $p(x) = L_{y'y'}$ and $q(x) = L_{yy} - \frac{d}{dx} L_{yy'}$. The ODE

$$A(u) = \frac{d}{dx} \left(p(x) \frac{du}{dx} \right) - q(x) u = 0, \; x \in [x, x_2]$$

is called the *Jacobi differential equation*. The $A(\cdot)$ differential operator is a self-adjoint *Sturm-Liouville* operator in the sense

$$u A(h) - h A(u) = \frac{d}{dx} \left(p(x)(uh' - hu') \right)$$

(*Lagrange identity*).

We return to the second variation, which we write in the form

$$I''(0) = -\int_{x_1}^{x_2} \frac{h}{u} uA(h)\,dx = -\int_{x_1}^{x_2} \frac{h}{u}\left(hA(u) + \frac{d}{dx}(p(x)(uh' - hu'))\right)dx.$$

In the last integral, we integrate by parts,

$$U = \frac{h}{u}, \quad dV = \frac{d}{dx}(p(x)(uh' - hu'))\,dx$$

$$dU = \frac{uh' - hu'}{u^2}\,dx, \quad V = p(x)(uh' - hu').$$

Setting $p = L_{y'y'}$, we deduce

$$I''(0) = \frac{h}{u}L_{y'y'}(uh' - hu')\Big|_{x_1}^{x_2} - \int_{x_1}^{x_2} \frac{h^2}{u}A(u)\,dx + \int_{x_1}^{x_2} L_{y'y'}\left(h' - h\frac{u'}{u}\right)^2 dx.$$

We add three conditions: (i) suppose $h(x_1) = 0$, $h(x_2) = 0$ (the first term disappears); (ii) assume that u is the solution of the Jacobi equation $A(u) = 0$, $x \in [x_1, x_2]$, $u(x_1) = 0$, $u'(x_1) \neq 0$ (the second term disappears); (iii) suppose $h' - h\frac{u'}{u} \neq 0$ (the sign of the third term is given by the sign of $L_{y'y'}$). It follows the

Sufficient conditions of Legendre-Jacobi:
(1) if $L_{y'y'} < 0$, $x \in [x_1, x_2]$, then the value $I(y)$ is maximum;
(2) if $L_{y'y'} > 0$, $x \in [x_1, x_2]$, then the value $I(y)$ is minimum.

2.3 Unitemporal Lagrangian Dynamics

Let us consider the simple integral functional

$$I(x(\cdot)) = \int_{t_0}^{t_1} L(t, x(t), \dot{x}(t))\,dt,$$

where $x(t) = (x^1(t), \ldots, x^n(t)) = (x^i(t))$, $i = \overline{1, n}$ is a collection of n functions (or equivalent, a function with values in $\mathbb{R}^n$, i.e., $x : [t_0, t_1] \to \mathbb{R}^n$) of class C^2, and $L = L(t, x, \dot{x})$ be a C^2 Lagrangian. We wish to extremize the functional $I(x(\cdot))$ with ends conditions $x(t_0) = x_0$ and $x(t_1) = x_1$.

Theorem 2.1: *A function $x(\cdot)$ of class C^2 which extremizes the functional I, necessarily satisfies the evolution Euler–Lagrange ODEs*

$$\frac{\partial L}{\partial x^i} - \frac{d}{dt}\frac{\partial L}{\partial \dot{x}^i} = 0$$

and the end conditions $x(t_0) = x_0$ and $x(t_1) = x_1$.

Here we have a system of n ODEs, usually of the second order, with n unknown functions $x^i(\cdot)$. The theorem shows that if the Euler–Lagrange system has solutions, then the minimizer (maximizer) of the functional I (assuming it exists) will be among the solutions. The solutions of the Euler–Lagrange system are called *extremals* or *critical points* of the Lagrangian L.

Proof. Let us consider that $x(t)$ is a solution of the previous problem. We build a variation around $x(t)$ of the form $x(t) + \varepsilon h(t)$, such that $h(t_0) = 0, h(t_1) = 0$. Here ε is a "small" parameter, and h is a "small" variation. The functional becomes an integral with parameter ε, i.e.,

$$I(\varepsilon) = \int_{x_0}^{x_1} L\left(t, x(t) + \varepsilon h(t), \dot{x}(t) + \varepsilon \dot{h}(t)\right) dt.$$

The necessary extremum condition is imposed,

$$0 = \frac{d}{d\varepsilon} I(\varepsilon)\bigg|_{\varepsilon=0} = (\cdots) = \int_{t_0}^{t_1} \left(\frac{\partial L}{\partial x^j} h^j + \frac{\partial L}{\partial \dot{x}^j} \dot{h}^j\right)(t) \, dt$$

$$= BT + \int_{t_0}^{t_1} \left(\frac{\partial L}{\partial x^j} - \frac{d}{dt} \frac{\partial L}{\partial \dot{x}^j}\right) h^j \, dt,$$

where

$$BT = \frac{\partial L}{\partial \dot{x}^j} h^j(t)\bigg|_{t=t_0}^{t=t_1}$$

are boundary terms, obtained via integration by parts. If the function $x(t)$ has fixed ends, then the terms BT disappear, because $h(t_0) = 0, h(t_1) = 0$. Taking the arbitrary h vector function, we find the system of differential equations in the theorem.

The dynamics described by second-order Euler–Lagrange ODEs is called *unitemporal Euler–Lagrange dynamics*. □

2.3.1 Null Lagrangians

Here we introduce an important notion of independent interest in the calculus of variations. We use the functional

$$J(x(\cdot)) = \int_a^b L(t, x(t), \dot{x}(t)) \, dt.$$

Definition 2.1: The Lagrangian $L(t, x(t), \dot{x}(t))$ is called a null Lagrangian iff the associated Euler–Lagrange ODE is identically zero.

As example, we consider the variational problem

$$J(x(\cdot)) = \int_0^\infty [t^2 x'(t) + 2tx(t)] \, dt.$$

The Lagrangian $L = t^2 x'(t) + 2tx(t)$ gives

$$\frac{\partial L}{\partial x} - \frac{d}{dt} \frac{\partial L}{\partial x'} = 2t - \frac{d}{dt}(t^2) \equiv 0.$$

Consequently, L is a null Lagrangian.

Theorem 2.2: *A function $L(x, u, p)$ defined for all $(x, u, p) \in \mathbb{R}^3$ is a null Lagrangian if and only if it is a total derivative, i.e.,*

$$L(x, u, u') = \frac{d}{dx} A(x, u) = \frac{\partial A}{\partial x} + u' \frac{\partial A}{\partial u}$$

for some function A that depends only on x, u. This statement extends automatically to $\mathbb{R}^{1+2n}$.

Proof. In Euler–Lagrange ODE, the only term involving u'' is the last one, and so if the left-hand side is to vanish for all possible functions $u(x)$, the coefficient of u'' must vanish, so $\frac{\partial^2 L}{\partial p^2} = 0$ which implies $L(x, u, p) = f(x, u)p + g(x, u)$. Coming back in Euler–Lagrange equations, we find

$$u' \frac{\partial f}{\partial u} + \frac{\partial g}{\partial u} - \frac{\partial f}{\partial x} - u' \frac{\partial f}{\partial u} = \frac{\partial g}{\partial u} - \frac{\partial f}{\partial x} = 0.$$

It follows $f = \frac{\partial A}{\partial u}, g = \frac{\partial A}{\partial x}$. $\qquad\qquad\square$

2.3.2 Invexity Test

Let us formulate a sufficient non-standard condition for the existence of the minimum of the simple integral functional

$$J(x(\cdot)) = \int_{t_1}^{t_2} L(t, x(t), \dot{x}(t)) \, dt,$$

where $t \in [t_1, t_2]$, L is a C^2 Lagrangian and $x : [t_1, t_2] \to \mathbb{R}^n$ is a C^2 function.

A function $x^*(t)$ is called a *critical point of the functional J* if it is a solution of the Euler–Lagrange system

$$\frac{\partial L}{\partial x^i} - \frac{d}{dt} \frac{\partial L}{\partial \dot{x}^i} = 0, \ i = \overline{1, n},$$

to which are added the conditions at the ends.

Definition 2.2: Let $x, x^* : [t_1, t_2] \to \mathbb{R}^n$ be two C^2 functions. If there exists a vectorial function $\eta(t, x, x^*, \dot{x}, \dot{x}^*)$, with n components and with $\eta = 0$, for $x(t) = x^*(t)$, such that

$$J(x(\cdot)) - J(x^*(\cdot)) \geq \int_{t_1}^{t_2} \left(\eta^i \frac{\partial L}{\partial x^{*i}} + \frac{d\eta^i}{dt} \frac{\partial L}{\partial \dot{x}^{*i}} \right) dt,$$

for any function $x(\cdot)$, then the functional J is called invex at the point $x^*(\cdot)$ on the interval $[t_1, t_2]$ with respect to the vectorial function η.

The functional J is called invex if it is invex at each point $x^*(\cdot)$.

Theorem 2.3: *The functional J is invex if and only if each critical point is a global minimum point.*

Proof. Suppose that the functional J is invex. Then, if $x^*(t)$ is a solution of Euler–Lagrange system, we have

$$J(x(\cdot)) - J(x^*(\cdot)) \geq \int_{t_1}^{t_2} \left(\frac{\partial L}{\partial x^{*i}} - \frac{d}{dt} \frac{\partial L}{\partial \dot{x}^{*i}} \right) \eta^i \, dt + \eta^i \frac{\partial L}{\partial \dot{x}^{*i}} \bigg|_{t_1}^{t_2} = 0.$$

It follows that a critical point is a global minimum point.

Conversely, suppose that each critical point is a global minimum point. If $x^*(t)$ is a critical point, then we set $\eta = 0$. If $x^*(t)$ is not a critical point, then

$$\frac{\partial L}{\partial x^{*i}} - \frac{d}{dt} \frac{\partial L}{\partial \dot{x}^{*i}} \neq 0$$

for at least one index i. The nonzero vector

$$\xi^i = \frac{\partial L}{\partial x^{*i}} - \frac{d}{dt} \frac{\partial L}{\partial \dot{x}^{*i}}$$

permits to define the vector

$$\eta^k = \frac{L(t, x, \dot{x}) - L(t, x^*, \dot{x}^*)}{2 \delta_{ij} \xi^i \xi^j} \xi^k.$$

The vector field $\eta = (\eta^k)$, thus defined, satisfies the relation from the invexity definition.

$\square$

2.4 Lavrentiev Phenomenon

Let us consider again the variational problem

$$\inf\{J(x(\cdot)) := \int_{t_0}^{t_1} L(t, x(t), \dot{x}(t)) \, dt \, : \, x(t_0) = a, \ x(t_1) = b\}.$$

For some concrete problems we may have some unpleasant surprises. Specifically, the Euler–Lagrange equation is written independently of the integration interval and the endpoint conditions. It may happen that the solutions of this equation are not defined on a portion of the integration interval, or that no solution satisfies the conditions at the ends.

Example 2.2: *The Lagrangian $L(t, x(t), \dot{x}(t)) = t^2 \dot{x}^2(t)$ produces the Euler–Lagrange equation $t^2 \dot{x} = c_1$, with the general solution $x(t) = c_2 - c_1/t$. Obviously the solution cannot be used on an interval that contains 0.*

In these situations it is natural to look for the extrema of the functional in a wider class of functions.

On the other hand, in order to obtain the necessary conditions, a class of (piecewise) smooth functions is the usual choice, whereas for existence theorems the much larger class of absolutely continuous functions is used (an absolutely continuous function is a continuous function which carries null sets in null sets; such a function is a.e. derivable and is an indefinite integral of its derivative). The Lavrentiev Phenomenon (1926) [37] occurs when the minimum value over the absolutely continuous functions was strictly less than the infimum over the smooth functions, or Lipschitz functions.

Problem 2.4.1: Find

$$\min\left\{\int_\tau^T t^2[\dot{x}(t)]^6 \, dt \; : \; x(\tau) = 0, \; x(T) = \alpha T^{1/3}\right\},$$

where $0 \le \tau < T$ and $\alpha > 0$ are given.

Solution. The Euler–Lagrange equation gives $x(t) = 5/3\,c\,t^{3/5} + d$, where the constants c and d are determined by the end conditions. The result is

$$x(t) = \alpha T^{1/3} \frac{t^{3/5} - \tau^{3/5}}{T^{3/5} - \tau^{3/5}}.$$

Consequently the minimum value of the problem is

$$\min \int_\tau^T t^2[\dot{x}(t)]^6 \, dt = \left(\frac{3}{5}\right)^5 \frac{\alpha^6 T^2}{(T^{3/5} - \tau^{3/5})^5}.$$

Note that for fixed α, the right-hand side tends to $+\infty$ as $T - \tau \to 0$. This observation is central to the developments below.

Let us now consider another basic problem.

Example 2.3: (*Mania's example – 1934*) *Let us now consider the basic optimum problem*

$$\min\left\{J(x(\cdot)) := \int_0^1 (x^3(t) - t)^2[\dot{x}(t)]^6 \, dt \; : \; x(0) = 0, \; x(1) = 1\right\}.$$

The Lagrangian $L(t, x, \dot{x}) = (x^3 - t)^2[\dot{x}(t)]^6$, which was introduced by B. Maniá in [40], is clearly nonnegative, so the absolutely continuous function $\bar{x}(t) = t^{1/3}$ minimizes J by giving $J(\bar{x}(\cdot)) = 0$. We will show that the values of J on $Lipp[0,1]$ form a set of positive numbers bounded away from zero, and thus verify LP.

For each $\alpha \in (0,1)$, define the set $R_\alpha = \{(t,x) : t \in (0,1), 0 \le x \le \alpha t^{1/3}\}$. In R_α, one has $L(t,x,v) \ge ((\alpha t^{1/3})^3 - t)^2 = (1 - \alpha^3)^2 t^2 v^6$. For any fixed α, any admissible arc $x \in Lipp$ with $x(0) = 0$ must spend some time in R_α. More precisely, let us define $0 \le \tau < T < 1$, where

$$\tau = \sup\{t \ge 0 : x(t) = 0\} \quad \text{and} \quad T = \inf\{t \ge \tau : x(t) = \alpha t^{1/3}\}.$$

From the analysis of the problem above, it follows that

$$J(x(\cdot)) \ge (1 - \alpha^3)^2 \int_\tau^T t^2[\dot{x}(t)]^6 \, dt$$

$$\ge \left(\frac{3}{5}\right)^5 (1 - \alpha^3)^2 \frac{\alpha^6 T^2}{(T^{3/5} - \tau^{3/5})^5} \ge \left(\frac{3}{5}\right)^5 (1 - \alpha^3)^2 \alpha^6$$

*(the last inequality holds because $\frac{T^2}{(T^{3/5}-\tau^{3/5})^5} \geq 1$). The maximum value of the right side occurs when $\alpha = 2^{1/3}$; we deduce that for every admissible $x \in Lipp$, we have $J(x(\cdot)) \geq \left(\frac{1}{2}\right)^4 \left(\frac{3}{5}\right)^5$.
This confirms LP. Note that $\bar{x}(t) = t^{1/3}$ is absolutely continuous, but not in $Lipp$ because its derivative in 0 is $+\infty$.*

Remark 2.1: The Euler–Lagrange equation for the Lagrangian $L(t, x, \dot{x}) = (x^3 - t)^2 [x'(t)]^6$ splits into one algebraic equation $x^3 - t = 0$ and the ODE

$$\ddot{x} + \frac{x^2}{x^3 - t} \dot{x}^2 - \frac{2}{x^3 - t} \dot{x} = 0.$$

So, $\bar{x}(t) = t^{1/3}$ is a singular solution for Euler–Lagrange equation, specifying only the initial conditions $x(0) = 0$, $x(1) = 1$. Second-order ODE may produce other minimizing extremals, possibly on other intervals and with other initial conditions.

The Lavrentiev Phenomenon identifies a difference in the infimum of a minimization problem across different classes of admissible functions. It is well known that any absolutely continuous function $\bar{x}$ is uniformly approximable by Lipschitz functions x_n. The previous example shows that sometimes $J(\bar{x}(\cdot)) < \lim J(x_n(\cdot))$. Hence a great warning for variational integrators because they always represent Lipschitzian functions.

Hilbert statement Every problem of the calculus of variations has a solution, provided that the word *solution* is suitably understood.

2.5 Unitemporal Hamiltonian Dynamics

Let us consider the C^2 Lagrangian $L(t, x(t), \dot{x}(t))$, where

$$t \in [t_0, t_1], \quad x = (x^1, ..., x^n) : [t_0, t_1] \to \mathbb{R}^n.$$

Now, for fixed $x(\cdot)$, we define the *generalized momentum*

$$p = (p_i), \quad p_i(t) = \frac{\partial L}{\partial \dot{x}^i}(t, x(t), \dot{x}(t)), \quad t \in [t_0, t_1].$$

Suppose that, for $\forall (x, p) \in \mathbb{R}^{2n}$, $t \in [t_0, t_1]$, this system defines the function $\dot{x} = \dot{x}(t, x, p)$. For this, locally, according to the implicit function theorem, it is necessary and sufficient that $\det\left(\frac{\partial^2 L}{\partial \dot{x}^i \partial \dot{x}^j}\right) \neq 0$. In this case, the Lagrangian L is called *regular* and enters in Legendrian duality with the *Hamiltonian*

$$H(t, x, p) = \dot{x}^i(t, x, p) \frac{\partial L}{\partial \dot{x}^i}(t, x, \dot{x}(t, x, p)) - L(t, x, \dot{x}(t, x, p)),$$

with summation over the index i. Shortly,

$$H = \dot{x}^i p_i - L.$$

Theorem 2.4: *If $x(\cdot)$ is a solution of the second-order system of Euler–Lagrange equations and the momentum p is defined as above, then the pair $(x(\cdot), p(\cdot))$ is a solution of the first-order Hamiltonian system*

$$\dot{x}^i(t) = \frac{\partial H}{\partial p_i}(t, x(t), p(t)), \quad \dot{p}_i(t) = -\frac{\partial H}{\partial x^i}(t, x(t), p(t)).$$

If L is autonomous (i.e., does not depend explicitly on t), then it follows that H is a first integral of the Hamiltonian ODEs (conservation law).

Proof. The result can be found either by direct variational calculation or by Euler–Lagrange equations (second-order!). Here, we prefer the second way.
  We find

$$\frac{\partial H}{\partial x^j} = p_i \frac{\partial \dot{x}^i}{\partial x^j} - \frac{\partial L}{\partial x^j} - \frac{\partial L}{\partial \dot{x}^i} \frac{\partial \dot{x}^i}{\partial x^j} = -\frac{\partial L}{\partial x^j}.$$

Then, we remark that Euler–Lagrange ODEs produce

$$\dot{p}_i(t) = \frac{\partial L}{\partial x^i}(t, x(t), \dot{x}(t)) = \frac{\partial L}{\partial x^i}(x(t), \dot{x}(t, x(t), p(t)))$$

$$= -\frac{\partial H}{\partial x^i}(t, x(t), p(t)).$$

Also

$$\frac{\partial H}{\partial p_j}(x, p) = \dot{x}^j(t, x, p) + p_i \frac{\partial \dot{x}^i}{\partial p_j} - \frac{\partial L}{\partial \dot{x}^i} \frac{\partial \dot{x}^i}{\partial p_j} = \dot{x}^j(t, x, p) = \dot{x}^j(t),$$

i.e.,

$$\dot{x}^j(t) = \frac{\partial H}{\partial p_j}(t, x(t), p(t)).$$

In conclusion, let's note that (total derivative = partial derivative)

$$\frac{dH}{dt} = \frac{\partial H}{\partial x^i}\frac{dx^i}{dt} - \frac{\partial H}{\partial p_i}\frac{dp_i}{dt} + \frac{\partial H}{\partial t} = \frac{\partial H}{\partial t}.$$

Some authors add the equation $\dfrac{dH}{dt} = \dfrac{\partial H}{\partial t}$ to Hamilton's equations, considering that t and H are dual variables. The autonomous case $L = L(x(t), \dot{x}(t))$ produces $H = H(x(t), p(t))$ and hence $\dfrac{dH}{dt} = 0$ (conservation law). □

Remark 2.2: Two Lagrangians L^1 and L^2 connected by the *gauge transformation*

$$L^2 = L^1 + \frac{d}{dt}f(x, t) = L^1 + \dot{x}^i\frac{\partial f}{\partial x^i} + \frac{\partial f}{\partial t}$$

produce the same Euler–Lagrange ODEs. The appropriate moments satisfy the relationship

$$p_i^2 = \frac{\partial L^2}{\partial \dot{x}^i} = \frac{\partial L^1}{\partial \dot{x}^i} + \frac{\partial f}{\partial x^i} = p_i^1 + \frac{\partial f}{\partial x^i}.$$

Example 2.4: *The classical Lagrangian in mechanics is*

$$L(x, \dot{x}) = \frac{1}{2} m ||\dot{x}||^2 - V(x)$$

(the difference between kinetic energy and potential energy). The Euler–Lagrange differential equation

$$m\ddot{x}(t) = -\nabla V(x(t))$$

represents the Newton law. If we introduce the momentum $p = mv$, $v = ||\dot{x}||$, then we can build the Hamiltonian

$$H(x, p) = \frac{1}{m}\langle p, p \rangle - L\left(x, \frac{p}{m}\right) = \frac{||p||^2}{2m} + V(x),$$

which represents the total energy. The Hamilton ODEs are

$$\dot{x}(t) = \frac{p(t)}{m}, \quad \dot{p}(t) = -\nabla V(x(t)).$$

2.6 Particular Euler–Lagrange ODEs

Let us consider again a C^2 Lagrangian $L(t, x(t), \dot{x}(t))$, $t \in \mathbb{R}$, $x(t) \in \mathbb{R}^n$ and the associated Euler–Lagrange ODEs system

$$\frac{\partial L}{\partial x^i} - \frac{d}{dt}\frac{\partial L}{\partial \dot{x}^i} = 0, \; i = \overline{1, n},$$

or explicitly

$$\frac{\partial L}{\partial x^i} - \frac{\partial^2 L}{\partial \dot{x}^i \partial t} - \frac{\partial^2 L}{\partial \dot{x}^i \partial x^j} \dot{x}^j(t) - \frac{\partial^2 L}{\partial \dot{x}^i \partial \dot{x}^j} \ddot{x}^j(t) = 0.$$

If the matrix of elements $g_{ij} = \frac{\partial^2 L}{\partial \dot{x}^i \partial \dot{x}^j}$ is nondegenerate, then the Euler–Lagrange ODEs system could be solved with respect to $\ddot{x}^j$. If the previous matrix is positive definite, we obtain a Newton Law $\ddot{x}^j = \dots$.

Absence of the variable t: In the autonomous case, the Euler–Lagrange ODEs system is reduced to

$$\frac{\partial L}{\partial x^i} - \frac{\partial^2 L}{\partial \dot{x}^i \partial x^j} \dot{x}^j(t) - \frac{\partial^2 L}{\partial \dot{x}^i \partial \dot{x}^j} \ddot{x}^j(t) = 0.$$

Contraction by $\dot{x}^i(t)$ gives

$$\frac{d}{dt}(\dot{x}^i(t) L_{\dot{x}^i(t)} - L) = 0.$$

We obtain a first integral $\dot{x}^i(t) L_{\dot{x}^i(t)} - L = c$, which means that the Hamiltonian $H = \dot{x}^i(t) L_{\dot{x}^i(t)} - L$ is conserved on solutions of Euler–Lagrange ODEs system.

Absence of the variable $x(t)$: The Euler–Lagrange ODEs system is $\frac{d}{dt}\frac{\partial L}{\partial \dot{x}^i} = 0$, i.e.,

$L_{\dot{x}^i(t)} = c_i$. The solution is of the form $\dot{x}^i(t) = g^i(t, c)$, i.e., $x^i(t) = \int_{t_0}^t g^i(t, c)\, dt$.

Absence of the velocity $\dot{x}(t)$: The Euler–Lagrange ODEs system becomes $L_{x(t)} = 0$. This is an algebraic equation. The solution $x = x(t)$ exists only if this curve happens to pass through the specified boundary points.

Linearity with respect to the velocity $\dot{x}(t)$: In this case the Lagrangian L could be written as $L(t, x(t), \dot{x}(t)) = a(t, x(t)) + b_i(t, x(t))\dot{x}^i(t)$, and the Euler–Lagrange ODEs system reduces to the algebraic equation $\dfrac{\partial a}{\partial x^i} - \dfrac{\partial b_i}{\partial t} = 0$. The variational problem is of no interest.

2.7 Multitemporal Lagrangian Dynamics

Traditionally, spatial coordinates and time usually play distinct roles: indexes associated with degrees of freedom, and the physical time in which systems evolve. This theory is satisfactory until we turn our attention to relativistically invariant equations (chiral fields, self-Gordon etc).

In relativity we use a 2-time $t = (t^1, t^2)$, where t^1 means intrinsic time and t^2 is the time of the observer. There are also a lot of problems where we have no reason to prefer one coordinate to another. This is why we refer to functions that depend on several time variables and that model multidimensional geometric evolutions. In this sense, by multi-time we mean a vector parameter of evolution.

Let t^α, $\alpha = \overline{1, m}$, be multi-time variables from *source space* $\mathbb{R}^m$ and x^i, $i = \overline{1, n}$, be the field variables in *target space* $\mathbb{R}^n$. We introduce the partial velocities $x_\alpha^i = \dfrac{\partial x^i}{\partial t^\alpha}$. All these determine the first-order jets fibration

$$J^1(\mathbb{R}^m, \mathbb{R}^n) = \{(t^\alpha, x^i, x_\alpha^i)\} \equiv \mathbb{R}^{m+n+mn}.$$

2.7.1 The Case of Multiple Integral Functionals

Let us consider a smooth *Lagrangian* $L(t, x(t), x_\gamma(t))$, $t \in \mathbb{R}_+^m$. We fix the multi-times $t_0, t_1 \in \mathbb{R}_+^m$ and two points $x_0, x_1 \in \mathbb{R}^n$. The parallelepiped $\Omega_{t_0, t_1} \subset \mathbb{R}_+^m$, fixed by opposite diagonal points $t_0 = (t_0^1, ..., t_0^m)$ and $t_1 = (t_1^1, ..., t_1^m)$, is equivalent to the closed interval $t_0 \leq t \leq t_1$, with respect to *partial order product* on $\mathbb{R}_+^m$. The classical problem of variational calculus requires us to find a C^2 m-sheet $x^*(\cdot) : \Omega_{t_0, t_1} \to \mathbb{R}^n$ which minimizes the *multiple integral functional*

$$I(x(\cdot)) = \int_{\Omega_{t_0, t_1}} L(t, x(t), x_\gamma(t)) \, dt^1 ... dt^m,$$

of all functions $x(\cdot)$ which satisfy either the boundary conditions $x(t_0) = x_0$, $x(t_1) = x_1$ or $x(t)|_{\partial \Omega_{t_0, t_1}}$ = given, using variation functions constrained by boundary conditions. The parallelepiped Ω can be replaced by a compact domain. The necessary conditions are contained in the following:

Theorem 2.5: *(Multitemporal Euler–Lagrange PDEs) Let D_γ be the total derivative operator with respect to t^γ. If the m-sheet $x^*(\cdot)$ minimizes the functional $I(x(\cdot))$ in previous sense, then $x^*(\cdot)$ is solution of multitemporal Euler–Lagrange PDEs*

$$\frac{\partial L}{\partial x^i} - D_\gamma \frac{\partial L}{\partial x^i_\gamma} = 0, \ i = \overline{1,n}, \ \gamma = \overline{1,m}, \qquad (E-L)_1$$

subject to boundary conditions.

Here we have a system of n PDEs, usually of second order, with n unknown functions $x^i(\cdot)$. The Theorem shows that if the system $(E-L)_1$ has solutions, then the minimizer of the multiple integral functional I (assuming it exists) will be among these solutions. The solutions of the system $(E-L)_1$ are called *extremals* or *critical points* of the Lagrangian L.

Proof. We consider that $x(t)$ is a solution of the previous problem. We build a variation of $x(t)$ under the form $x(t) + \varepsilon h(t)$, with $h|_{\partial \Omega_{t_0,t_1}} = 0$. Here ε is a "small" parameter, and h is a "small" variation. The functional becomes a function of ε, i.e., an integral with one parameter,

$$I(\varepsilon) = \int_{\Omega_{t_0,t_1}} L\left(t, x(t) + \varepsilon h(t), x_\gamma(t) + \varepsilon h_\gamma(t)\right) dt^1 ... dt^m.$$

The necessary extremum condition is imposed,

$$0 = \frac{d}{d\varepsilon} I(\varepsilon)\Big|_{\varepsilon=0} = (\cdots) = \int_{\Omega_{t_0,t_1}} \left(\frac{\partial L}{\partial x^j} h^j + \frac{\partial L}{\partial x^j_\gamma} h^j_\gamma\right)(t)\, dt^1 ... dt^m$$

$$= \text{BT} + \int_{\Omega_{t_0,t_1}} \left(\frac{\partial L}{\partial x^j} - D_\gamma \frac{\partial L}{\partial x^j_\gamma}\right) h^j \, dt^1 ... dt^m,$$

where D_γ is the total derivative operator, acting according to the rule (divergence operator)

$$D_\gamma \left(\frac{\partial L}{\partial x^j_\gamma} h^j\right) = h^j D_\gamma \left(\frac{\partial L}{\partial x^j_\gamma}\right) + \frac{\partial L}{\partial x^j_\gamma} D_\gamma h^j,$$

with sum over $\gamma = \overline{1,m}$, respectively $j = \overline{1,n}$. The term

$$\text{BT} = \int_{\Omega_{t_0,t_1}} D_\gamma \left(\frac{\partial L}{\partial x^j_\gamma} h^j\right) dt^1 ... dt^m = \int_{\partial \Omega_{t_0,t_1}} \delta_{\alpha\beta} \frac{\partial L}{\partial x^j_\alpha} h^j n^\beta(t) d\sigma$$

obtained by the divergence formula, using the unit vector $n^\beta(t)$ normal to the boundary $\partial \Omega_{t_0,t_1}$, contains data on boundary. The term BT disappears, since $h|_{\partial \Omega_{t_0,t_1}} = 0$. Arbitrarily taking the vector function h, we find the Euler–Lagrange PDEs system in the Theorem. $\square$

The dynamics described by second-order Euler–Lagrange PDEs are called *multitemporal Euler–Lagrange dynamics.*

2.7.2 Invexity Test

Let us formulate a sufficient condition for the existence of minimum of a multiple integral functional

$$I(x(\cdot)) = \int_{\Omega_{t_0,t_1}} L(t, x(t), x_\gamma(t)) \, dt^1 ... dt^m,$$

where $t \in \Omega_{t_0,t_1}$, L is a C^2 Lagrangian and $x : \Omega_{t_0,t_1} \to \mathbb{R}^n$ is a C^2 function.

A function $x^*(t)$ is called *critical point of the functional I* if it is solution of Euler–Lagrange PDEs system

$$\frac{\partial L}{\partial x^i} - D_\gamma \frac{\partial L}{\partial x^i_\gamma} = 0,$$

to which the boundary condition is added.

Definition 2.3: Let $x, x^* : \Omega_{t_0,t_1} \to \mathbb{R}^n$ be C^2 functions. If there exists a vectorial function $\eta(t, x, x^*, \dot{x}, \dot{x}^*)$, with n components and with $\eta = 0$, for $x(t) = x^*(t)$, such that

$$I(x(\cdot)) - I(x^*(\cdot)) \geq \int_{\Omega_{t_0,t_1}} \left(\eta^i \frac{\partial L}{\partial x^{*i}} + (D_\alpha \eta^i) \frac{\partial L}{\partial x^{*i}_\alpha} \right) dt^1 ... dt^m,$$

for any function $x(\cdot)$, then the functional I is called invex at the point $x^*(t)$ on the interval $[t_1, t_2]$ with respect to the vectorial function η.

The functional I is called invex if it is invex at any point $x^*(\cdot)$.

Theorem 2.6: *The functional I is invex if and only if each critical point is a global minimum point.*

Proof. Suppose that the functional I is invex. Then, if $x^*(t)$ is a solution of Euler–Lagrange system, we have

$$I(x(\cdot)) - I(x^*(\cdot)) \geq \int_{\Omega_{t_0,t_1}} \left(\frac{\partial L}{\partial x^{*i}} - D_\alpha \frac{\partial L}{\partial x^{*i}_\alpha} \right) \eta^i \, dt^1 ... dt^m$$

$$+ \int_{\partial \Omega_{t_0,t_1}} \eta^i \frac{\partial L}{\partial x^{*i}_\alpha} n^\alpha do = 0.$$

It follows that a critical point is a global minimum point.

Conversely, suppose that each critical point is a global minimum point. If $x^*(t)$ is a critical point, then we set $\eta = 0$. If $x^*(t)$ is not a critical point, then

$$\frac{\partial L}{\partial x^{*i}} - D_\alpha \frac{\partial L}{\partial x^{*i}_\alpha} \neq 0,$$

for at least one index i. The nonzero vector field

$$\xi^i = \frac{\partial L}{\partial x^{*i}} - D_\alpha \frac{\partial L}{\partial x^{*i}_\alpha}$$

permits to define the vector field

$$\eta^k = \frac{L(t,x,x_\alpha) - L(t,x^*,x_\alpha^*)}{2\delta_{ij}\xi^i\xi^j}\, \xi^k.$$

The vector field η, thus defined, satisfies the relation from the invexity definition. □

2.7.3 The Case of Path-Independent Curvilinear Integral Functionals

A smooth *Lagrangian* $L(t,x(t),x_\gamma(t))$, $t \in \mathbb{R}_+^m$ produces two smooth closed 1-forms (completely integrable):

- the differential

$$dL = \frac{\partial L}{\partial t^\gamma}\, dt^\gamma + \frac{\partial L}{\partial x^i}\, dx^i + \frac{\partial L}{\partial x_\gamma^i}\, dx_\gamma^i$$

of components $\left(\dfrac{\partial L}{\partial t^\gamma}, \dfrac{\partial L}{\partial x^i}, \dfrac{\partial L}{\partial x_\gamma^i}\right)$, with respect to the basis $(dt^\gamma, dx^i, dx_\gamma^i)$;

- the restriction of dL to $(t,x(t),x_\gamma(t))$, i.e., the pullback

$$dL|_{(t,x(t),x_\gamma(t))} = \left(\frac{\partial L}{\partial t^\beta} + \frac{\partial L}{\partial x^i}\frac{\partial x^i}{\partial t^\beta} + \frac{\partial L}{\partial x_\gamma^i}\frac{\partial x_\gamma^i}{\partial t^\beta}\right) dt^\beta,$$

of components (which contain partial accelerations)

$$D_\beta L = \frac{\partial L}{\partial t^\beta}(t,x(t),x_\gamma(t)) + \frac{\partial L}{\partial x^i}(t,x(t),x_\gamma(t))\frac{\partial x^i}{\partial t^\beta}(t)$$

$$+ \frac{\partial L}{\partial x_\gamma^i}(t,x(t),x_\gamma(t))\frac{\partial x_\gamma^i}{\partial t^\beta}(t),$$

with respect to the basis dt^β.

Let $L_\beta(t,x(t),x_\gamma(t))\,dt^\beta$ be closed Lagrange 1-form (completely integrable), i.e., $D_\beta L_\alpha = D_\alpha L_\beta$ or explicit

$$\frac{\partial L_\beta}{\partial t^\alpha} + \frac{\partial L_\beta}{\partial x^i}\frac{\partial x^i}{\partial t^\alpha} + \frac{\partial L_\beta}{\partial x_\gamma^i}\frac{\partial x_\gamma^i}{\partial t^\alpha} = \frac{\partial L_\alpha}{\partial t^\beta} + \frac{\partial L_\alpha}{\partial x^i}\frac{\partial x^i}{\partial t^\beta} + \frac{\partial L_\alpha}{\partial x_\gamma^i}\frac{\partial x_\gamma^i}{\partial t^\beta}.$$

If there exists a Lagrangian $L(t,x(t),x_\gamma(t))$ with the property $D_\beta L = L_\beta$ (the previous pull-back is a given closed 1-form), then the function $x(t)$ is a solution of the complete integrable PDEs system (of second order)

$$\frac{\partial L}{\partial t^\beta} + \frac{\partial L}{\partial x^i}\frac{\partial x^i}{\partial t^\beta} + \frac{\partial L}{\partial x_\gamma^i}\frac{\partial x_\gamma^i}{\partial t^\beta} = L_\beta.$$

Let Γ_{t_0,t_1} be an arbitrary piecewise C^1 curve joining the points t_0 and t_1. Let us introduce a new variational calculus problem asking to find an m-sheet $x^*(\cdot) : \Omega_{t_0,t_1} \to \mathbb{R}^n$ of class C^2 which minimizes the *path-independent curvilinear integral functional (action)*

$$J(x(\cdot)) = \int_{\Gamma_{t_0,t_1}} L_\beta(t,x(t),x_\gamma(t))\,dt^\beta, \quad \text{sum over } \beta = \overline{1,m},$$

of all functions $x(\cdot)$ which satisfy the boundary conditions $x(t_0) = x_0, x(t_1) = x_1$ or $x(t)|_{\partial\Omega_{t_0,t_1}}$ = given, using variation functions constrained by boundary conditions and by closing conditions (complete integrability) of the Lagrange 1-form. The parallelepiped Ω can be replaced by a compact domain.

Problem 2.7.1: (Fundamental problem) How can we characterize the function $x^*(\cdot)$ which solves the variational problem associated to the previous curvilinear functional?

Theorem 2.7: *Suppose there exists a Lagrangian $L(t, x(t), x_\gamma(t))$ with the property $D_\beta L = L_\beta$.*

(i) If the m-sheet $x^*(\cdot)$ is an extremal of L, then it is also an extremal of the differential dL.
(ii) If the m-sheet $x^*(\cdot)$ minimizes the functional $J(x(\cdot))$, then $x^*(\cdot)$ is a solution of multi-time PDEs system

$$\frac{\partial L}{\partial x^i} - D_\gamma \frac{\partial L}{\partial x^i_\gamma} = a_i, \ i = \overline{1,n}, \ \gamma = \overline{1,m}, \tag{$E-L$}_2$$

(sum over γ) which satisfies the boundary conditions, where a_i are arbitrary constants.

The second part of the Theorem shows that if the system $(E-L)_2$ has solutions, then the minimizer of the functional J (supposing that it exists) is among these solutions.

Proof. 1) The differential switches with the partial derivation operators,

$$0 = d\left(\frac{\partial L}{\partial x^i} - D_\gamma \frac{\partial L}{\partial x^i_\gamma}\right) = \frac{\partial (dL)}{\partial x^i} - D_\gamma \frac{\partial (dL)}{\partial x^i_\gamma}.$$

On components, this means

$$\frac{\partial}{\partial x^i}\left(\frac{\partial L}{\partial x^j}\right) - D_\gamma \frac{\partial}{\partial x^i_\gamma}\left(\frac{\partial L}{\partial x^j}\right) = 0$$

$$\frac{\partial}{\partial x^i}\left(\frac{\partial L}{\partial x^j_\alpha}\right) - D_\gamma \frac{\partial}{\partial x^i_\gamma}\left(\frac{\partial L}{\partial x^j_\alpha}\right) = 0$$

$$\frac{\partial}{\partial x^i}\left(\frac{\partial L}{\partial t^\alpha}\right) - D_\gamma \frac{\partial}{\partial x^i_\gamma}\left(\frac{\partial L}{\partial t^\alpha}\right) = 0.$$

2) The result is obtained by the equalities

$$0 = \frac{\partial (dL)}{\partial x^i} - D_\gamma \frac{\partial (dL)}{\partial x^i_\gamma} = d\left(\frac{\partial L}{\partial x^i} - D_\gamma \frac{\partial L}{\partial x^i_\gamma}\right). \qquad \square$$

Theorem 2.8: *If the m-sheet $x^*(\cdot)$ minimizes the functional $J(x(\cdot))$, then $x^*(\cdot)$ is a solution of multi-time PDEs system*

$$\frac{\partial L_\beta}{\partial x^i} - D_\gamma \frac{\partial L_\beta}{\partial x^i_\gamma} = 0, \ i = \overline{1, n}; \ \beta, \gamma = \overline{1, m}, \qquad (E - L)_3$$

which satisfies the boundary conditions.

The Theorem shows that if the PDEs system $(E - L)_3$ has solutions, then the minimizer of the functional J (supposing there exists) is among these solutions.

Proof. We consider that $x(t)$ is a solution of the previous problem. Implicitly, it must satisfy the complete integrability conditions of the Lagrangian 1-form $L_\beta(t, x(t), x_\gamma(t)) \, dt^\beta$, i.e.,

$$\frac{\partial L_\beta}{\partial t^\alpha} + \frac{\partial L_\beta}{\partial x^i} \frac{\partial x^i}{\partial t^\alpha} + \frac{\partial L_\beta}{\partial x^i_\gamma} \frac{\partial x^i_\gamma}{\partial t^\alpha} = \frac{\partial L_\alpha}{\partial t^\beta} + \frac{\partial L_\alpha}{\partial x^i} \frac{\partial x^i}{\partial t^\beta} + \frac{\partial L_\alpha}{\partial x^i_\gamma} \frac{\partial x^i_\gamma}{\partial t^\beta}.$$

We build a variation $x(t) + \varepsilon h(t)$, with $h(t_0) = 0, h(t_1) = 0$, of the function $x(t)$. Here ε is a "small" parameter, and h is a "small" variation. The functional becomes a function of ε, that is an integral with parameter,

$$J(\varepsilon) = \int_{\Gamma_{t_0, t_1}} L_\beta \left(t, x(t) + \varepsilon h(t), x_\gamma(t) + \varepsilon h_\gamma(t) \right) dt^\beta.$$

We accept that the variation h satisfies the complete integrability conditions of the 1-form

$$L_\beta \left(t, x(t) + \varepsilon h(t), x_\gamma(t) + \varepsilon h_\gamma(t) \right) dt^\beta.$$

This condition adds a new PDE

$$\frac{\partial L_\beta}{\partial x^i} \frac{\partial h^i}{\partial t^\alpha} + \frac{\partial L_\beta}{\partial x^i_\gamma} \frac{\partial h^i_\gamma}{\partial t^\alpha} = \frac{\partial L_\alpha}{\partial x^i} \frac{\partial h^i}{\partial t^\beta} + \frac{\partial L_\alpha}{\partial x^i_\gamma} \frac{\partial h^i_\gamma}{\partial t^\beta},$$

which shows that the set of functions $h(t)$ is a vector space, and the set of functions $x(t) + \varepsilon h(t)$ is an affine space. It is required

$$0 = \frac{d}{d\varepsilon} J(\varepsilon) \Big|_{\varepsilon=0} = (\cdots) = \int_{\Gamma_{t_0, t_1}} \left(\frac{\partial L_\beta}{\partial x^j} h^j + \frac{\partial L_\beta}{\partial x^j_\gamma} h^j_\gamma \right) dt^\beta$$

$$= BT + \int_{\Gamma_{t_0, t_1}} \left(\frac{\partial L_\beta}{\partial x^j} - D_\gamma \frac{\partial L_\beta}{\partial x^j_\gamma} \right) h^j \, dt^\beta,$$

where D_γ is the total derivative operator, which acts as a divergence

$$D_\gamma \left(\frac{\partial L_\beta}{\partial x^j_\gamma} h^j \right) = h^j D_\gamma \left(\frac{\partial L_\beta}{\partial x^j_\gamma} \right) + \frac{\partial L_\beta}{\partial x^j_\gamma} D_\gamma h^j.$$

We use variations h null at the ends for which

$$D_\gamma \left(h^i \frac{\partial L_\beta}{\partial x^i_\gamma} \right) = D_\beta \left(h^i \frac{\partial L_\gamma}{\partial x^i_\gamma} \right).$$

This restricts the vector space of the variations $h(t)$ to a subspace. Then the terms on the boundary are written

$$BT = \int_{\Gamma_{t_0,t_1}} D_\gamma \left(\frac{\partial L_\beta}{\partial x^j_\gamma} h^j \right) dt^\beta$$

$$= \int_{\Gamma_{t_0,t_1}} D_\beta \left(\frac{\partial L_\gamma}{\partial x^j_\gamma} h^j \right) dt^\beta = \frac{\partial L_\gamma}{\partial x^j_\gamma} h^j |_{t_0}^{t_1}.$$

The terms BT vanish since $h(t_0) = 0, h(t_1) = 0$. It rests

$$0 = \int_{\Gamma_{t_0,t_1}} \left(\frac{\partial L_\beta}{\partial x^j} - D_\gamma \frac{\partial L_\beta}{\partial x^j_\gamma} \right) h^j \, dt^\beta.$$

Since the curve Γ_{t_0,t_1} is arbitrary, we find the PDEs system in the Theorem. $\qquad \square$

The dynamics described by the second-order Euler–Lagrange PDEs is called *multitemporal Euler–Lagrange dynamics*.

2.7.4 Invexity Test

Let us now formulate a sufficiently unorthodox condition for the existence of the functional minimum

$$J(x(\cdot)) = \int_{\Gamma_{t_0,t_1}} L_\beta(t, x(t), x_\gamma(t)) \, dt^\beta = \text{minimum},$$

where $\beta = \overline{1,m}$, $t \in \Omega_{t_0,t_1}$, $L_\beta(t, x(t), x_\gamma(t)) \, dt^\beta$ is a Lagrangian 1-form of class C^2 and $x : \Omega_{t_0,t_1} \to \mathbb{R}^n$ is a C^2 function.

A function $x^*(t)$ is called *critical point of the functional J* if it is a solution of Euler–Lagrange PDEs system

$$\frac{\partial L_\alpha}{\partial x^i} - D_\gamma \frac{\partial L_\alpha}{\partial x^i_\gamma} = 0,$$

to which are added the conditions at the boundary.

Definition 2.4: Let $x, x^* : \Omega_{t_0,t_1} \to \mathbb{R}^n$ be C^2 functions. If there exists a vectorial function $\eta(t, x, x^*, \dot{x}, \dot{x}^*)$, with n components and with $\eta = 0$, for $x(t) = x^*(t)$, such that

$$J(x(\cdot)) - J(x^*(\cdot)) \geq \int_{\Gamma_{t_0,t_1}} \left(\eta^i \frac{\partial L_\alpha}{\partial x^{*i}} + (D_\beta \eta^i) \frac{\partial L_\alpha}{\partial x^{*i}_\beta} \right) dt^\alpha,$$

for any function $x(\cdot)$ and for any curve Γ_{t_0,t_1}, then J is called invex at the point $x^*(\cdot)$ on the interval Ω_{t_0,t_1} with respect to the vectorial function η.

The functional J is called invex if it is invex at any point $x^*(\cdot)$.

Theorem 2.9: *The functional J is invex if and only if any critical point is a global minimum point.*

Proof. Suppose that the functional J is invex. Then, if $x^*(t)$ is a solution of Euler–Lagrange system, we have

$$J(x(\cdot)) - J(x^*(\cdot)) \geq \int_{\Gamma_{t_0,t_1}} \left(\frac{\partial L_\alpha}{\partial x^{*i}} - D_\beta \frac{\partial L_\alpha}{\partial x^{*i}_\beta} \right) \eta^i \, dt^\alpha$$

$$+ \int_{\Gamma_{t_0,t_1}} D_\alpha \left(\eta^i \frac{\partial L_\beta}{\partial x^{*i}_\beta} \right) dt^\alpha = 0.$$

It follows that a critical point is a global minimum point.

Conversely, suppose that any critical point is a global minimum point. If $x^*(t)$ is a critical point, then we set $\eta = 0$. If $x^*(t)$ is not a critical point, then $\exists \alpha \in \{1, ..., m\}$, $\exists i \in \{1, ..., n\}$, let say $\alpha = 1$, $i = 1$, for which

$$\frac{\partial L_1}{\partial x^{*1}} - D_\beta \frac{\partial L_1}{\partial x^{*1}_\beta} \neq 0.$$

The vector field ξ of components

$$\xi^1 = \frac{\partial L_1}{\partial x^{*1}} - D_\beta \frac{\partial L_1}{\partial x^{*1}_\beta}, \quad \xi^i = 0, \, i \neq 1$$

permits to define the vector field η of components

$$\eta^1 = \frac{L_1(t, x, \dot{x}) - L_1(t, x^*, \dot{x}^*)}{2\delta_{ij}\xi^i\xi^j} \xi^1, \quad \eta^i = 0, \, i \neq 1.$$

The vector field η, with previous components, satisfies the relation from the definition of invexity. $\qquad \square$

2.8 Multitemporal Hamiltonian Dynamics

In the multitemporal case, significant Hamiltonian dynamics refers to multiple integral functionals. Let us consider the hyper-parallelepiped $\Omega_{t_0,t_1} \subset \mathbb{R}^m$ determined by diagonal opposite points t_0, t_1 of $\mathbb{R}^m$. If on $\mathbb{R}^m$ is considered *partial order product*, then Ω_{t_0,t_1} is identified to the interval $[t_0, t_1]$. Let us consider the C^2 Lagrangian $L(t, x(t), x_\gamma(t))$, where

$$t = (t^1, ..., t^m) = (t^\alpha) \in \Omega_{t_0,t_1}, \quad x = (x^1, ..., x^n) : \Omega_{t_0,t_1} \to \mathbb{R}^n.$$

Now, for fixed $x(\cdot)$, we define the *generalized multi-momentum*

$$p = (p_i^\alpha), \quad p_i^\alpha(t) = \frac{\partial L}{\partial x_\alpha^i}(t, x(t), x_\gamma(t)), \quad t \in \Omega_{t_0, t_1}.$$

Suppose that, for $\forall x \in \mathbb{R}^n$, $p \in \mathbb{R}^{mn}$, $t \in \Omega_{t_0, t_1}$, this system defines the functions $x_\gamma^i = x_\gamma^i(x, p, t)$. In other words, locally, according the implicit function theorem, it is necessary and sufficient to have $\det\left(\dfrac{\partial^2 L}{\partial x_\alpha^i \partial x_\beta^j}\right) \neq 0$, where the pairs (α, i) mean lines, and the pairs (β, j) mean columns. In this case the Lagrangian L is called *regular* and enters in *Legendrian duality* with the *Hamiltonian* (sum over the indices i and α)

$$H(t, x, p) = x_\alpha^i(t, x, p) \frac{\partial L}{\partial x_\alpha^i}(t, x, x_\gamma(t, x, p)) - L(t, x, x_\gamma(t, x, p))$$

or shortly

$$H(t, x, p) = x_\alpha^i(t, x, p) \, p_i^\alpha(t) - L(t, x, p).$$

In this context we can introduce the *energy-impulse tensor T* of components

$$T_\beta^\alpha(t, x, p) = x_\beta^i(t, x, p) \, p_i^\alpha(t) - L(t, x, p) \, \delta_\beta^\alpha$$

and the *Hamilton tensor field* of components

$$H_\beta^\alpha(t, x, p) = x_\beta^i(t, x, p) \, p_i^\alpha(t) - \frac{1}{m} L(t, x, p) \, \delta_\beta^\alpha.$$

The trace H_α^α (sum over α) is the Hamiltonian.

The energy-impulse tensor is a physical quantity that describes the density and flux of energy and momentum in space-time, generalizing the stress tensor of Newtonian physics. It is an attribute of matter, radiation and non-gravitational force fields.

Theorem 2.10: *If $x(\cdot)$ is a solution of second-order Euler–Lagrange PDEs and the multi-momentum p is defined as above, then the pair $(x(\cdot), p(\cdot))$ is a solution of first-order Hamilton PDEs*

$$\frac{\partial x^i}{\partial t^\beta}(t) = \frac{\partial H}{\partial p_i^\beta}(t, x(t), p(t)), \quad \frac{\partial p_i^\alpha}{\partial t^\alpha}(t) = -\frac{\partial H}{\partial x^i}(t, x(t), p(t)),$$

with sum over α. Moreover Div $T = 0$, i.e., $D_\alpha T_\beta^\alpha = 0$ on solutions of the Hamiltonian PDEs system (zero total divergence, conservation law).

Proof. The result can be found either by direct variational calculus or by Euler–Lagrange PDEs. Here, we prefer the second way.

By calculation we find

$$\frac{\partial H}{\partial x^j}(t, x, p) = -\frac{\partial L}{\partial x^j}(t, x, p).$$

Also

$$p_i^\alpha(t) = \frac{\partial L}{\partial x_\alpha^i}(t, x(t), x_\gamma(t))$$

if and only if

$$\frac{\partial x^i}{\partial t^\alpha}(t) = x_\alpha(t, x(t), p(t)).$$

That is why the Euler–Lagrange PDEs produce

$$\frac{\partial p_i^\alpha}{\partial t^\alpha}(t) = \frac{\partial L}{\partial x^i}(t, x(t), x_\gamma(t)) = -\frac{\partial H}{\partial x^i}(t, x(t), p(t)).$$

Also

$$\frac{\partial H}{\partial p_j^\alpha}(x, p, t) = x_\alpha^j(t, x, p),$$

i.e.,

$$\frac{\partial x^j}{\partial t^\alpha}(t) = \frac{\partial H}{\partial p_j^\alpha}(t, x(t), p(t)).$$

In the case of several evolutionary variables, the Hamiltonian H is not preserved, i.e., $D_\alpha H \neq 0$, even in the autonomous case. Instead, the conservation law says that the divergence of the energy-impulse tensor T_β^α is zero, that is, $D_\alpha T_\beta^\alpha = 0$ (sum over α) on solutions of the Hamiltonian PDEs system. □

2.9 Particular Euler–Lagrange PDEs

Let us consider a C^2 Lagrangian $L(t, x(t), x_\gamma(t))$, $t \in \mathbb{R}^m$, $x(t) \in \mathbb{R}^n$ and the associated Euler–Lagrange PDEs system

$$\frac{\partial L}{\partial x^i} - \frac{\partial}{\partial t^\gamma}\frac{\partial L}{\partial x_\gamma^i} = 0, \ i = \overline{1, n},$$

with sum over $\gamma = \overline{1, m}$. The last term $\frac{\partial}{\partial t^\gamma}\frac{\partial L}{\partial x_\gamma^i}$ is of divergence type. The Euler–Lagrange system can be written in detail

$$\frac{\partial L}{\partial x^i} - \frac{\partial^2 L}{\partial x_\gamma^i \partial t^\gamma} - \frac{\partial^2 L}{\partial x_\gamma^i \partial x^j} x_\gamma^j(t) - \frac{\partial^2 L}{\partial x_\gamma^i \partial x_\lambda^j} x_{\gamma\lambda}^j(t) = 0,$$

highlighting unknown $x(t)$. If the matrix of elements $G_{ij}^{\gamma\lambda} = \frac{\partial^2 L}{\partial x_\gamma^i \partial x_\lambda^j}$, (γ, i) – lines, (λ, j) – columns, is nondegenerate, then the momentum $p_i^\alpha(t)$ is well defined via the algebraic system

$$p = (p_i^\alpha), \ p_i^\alpha(t) = \frac{\partial L}{\partial x_\alpha^i}(t, x(t), x_\gamma(t)), \ t \in \Omega_{t_0, t_1}.$$

If in addition the matrix $\left(G_{ij}^{\gamma\lambda}\right)$ is also decomposable, i.e., $G_{ij}^{\gamma\lambda} = h^{\gamma\lambda}g_{ij}$, then the Euler–Lagrange PDEs system could be solved in the form $h^{\gamma\lambda}x_{\gamma\lambda}^j = \dots$. If the previous matrix is positive definite, we obtain a Newton Law, i.e., a Poisson system.

Absence of the variable t: In the autonomous case, the Euler–Lagrange PDEs system is reduced to

$$\frac{\partial L}{\partial x^i} - \frac{\partial^2 L}{\partial x_\gamma^i \partial x^j} x_\gamma^j(t) - \frac{\partial^2 L}{\partial x_\gamma^i \partial x_\lambda^j} x_{\gamma\lambda}^j(t) = 0.$$

In the multitemporal case, although the Lagrangian is autonomous, the Hamiltonian is not conserved.

Absence of the variable $x(t)$: The Euler–Lagrange PDEs system is $\dfrac{\partial}{\partial t^\gamma}\dfrac{\partial L}{\partial x_\gamma^i} = 0$, i.e., a divergence type system.

Absence of the partial velocity $x_\gamma(t)$: The Euler–Lagrange PDEs system becomes $L_{x(t)} = 0$. This is an algebraic equation. The solution $x = x(t)$ exists only if this m-sheet happens to pass through the specified boundary condition.

Linearity with respect to the partial velocity $x_\gamma(t)$: In this case the Lagrangian L could be written as $L(t, x(t), x_\lambda(t)) = A(t, x(t)) + B_i^\lambda(t, x(t))x_\lambda^i(t)$, and the Euler–Lagrange PDEs system reduces to the algebraic equation $\dfrac{\partial A}{\partial x^i} - \dfrac{\partial B_i^\lambda}{\partial t^\lambda} = 0$, with unknown $x(t)$. The associated variational problem is of no interest.

Conservation of energy-impulse tensor: The *energy-impulse tensor* T of components

$$T_\beta^\alpha(t, x, p) = x_\beta^i(t, x, p) \ p_i^\alpha(t) - L(t, x, p) \ \delta_\beta^\alpha$$

is conserved in the sense that its divergence vanishes, i.e., $D_\alpha T_\beta^\alpha = 0$ along the solutions of Euler–Lagrange PDEs system (sum over α).

2.10   Maple Application Topics

Problem 2.10.1: Determine the values of the simple integral functional $I(y) = \displaystyle\int_0^1 y'(x)^2 \, dx$, for $y(x) = x$ and $y(x) = x^3$.

Problem 2.10.2: Consider the functional

$$I(y) = \int_1^2 xy'(x)^2 \, dx, \ y(1) = 0, y(2) = 1.$$

(a) Show that the extremal path satisfies $xy'(x) = \text{constant}$.
(b) Hence show that the extremal function is $y(x) = \ln x / \ln 2$.

Problem 2.10.3: **(Fermat's principle)** This principle states that light takes a path that (locally) minimizes the optical length between its endpoints. If the x-coordinate is chosen as the parameter, and $y = f(x)$ along the path, then the optical length is given by

$$I(f) = \int_{x_0}^{x_1} n(x, f(x))\sqrt{1 + f'(x)^2}\, dx,$$

where the refractive index $n(x, y)$ depends upon the material. Find Euler–Lagrange ODE.

Problem 2.10.4: **(A problem in navigation)** Consider a river with straight, parallel banks at distance a, and a boat that can travel with constant speed c in still water. The problem is to cross the river in the shortest time, starting and landing at given points, when there is a current.

We choose the y-axis to be the left bank, the line $x = a$ to be the right bank, and the starting point to be the origin. The water is assumed to be moving parallel to the banks with speed $v(x)$, a known function of the distance from the left bank. Then the time of passage along the path $y(x)$ from the point $(0, 0)$ on the left bank to the point (a, A) on the right bank, assuming that $c > max(v(x))$, can be shown to be the functional

$$T(y) = \int_0^a \frac{\sqrt{c^2(1 + y'(x)^2) - v(x)^2} - v(x)y'(x)}{c^2 - v^2(x)}\, dx, \quad y(0) = 0, y(a) = A.$$

This functional maps paths $y(x)$ to the times to cross the river and depends explicitly on the speed profile $v(x)$.

The problem is to find paths that minimize the functional $T(y)$.

Problem 2.10.5: **(Central field motion)** Consider the Lagrangian

$$L(x, y, \dot{x}, \dot{y}) = \frac{1}{2}(\dot{x}^2 + \dot{y}^2) - U\left(\sqrt{x^2 + y^2}\right)$$

of a particle in the plane subjected to a radial potential field. Change the Cartesian coordinates (x, y) into polar coordinates (r, θ) and write the Euler–Lagrange ODEs.

Solution. We start with $x = r\cos\theta$, $y = r\sin\theta$. It follows

$$L(r, \theta, \dot{r}, \dot{\theta}) = \frac{1}{2}(r^2\dot{\theta}^2 + \dot{r}^2) - U(r).$$

The Euler–Lagrange ODEs are

$$\frac{d}{dt}r^2\dot{\theta} = 0, \quad \frac{d}{dt}\dot{r} = -U'(r) + r\dot{\theta}^2.$$

Missing t, the Hamiltonian H is a conserved quantity.

Problem 2.10.6: Show that the critical points (p, x) of the functional

$$\int_0^T \left[\frac{1}{2}(p_i \dot{x}^i - x^i \dot{p}_i) + H(p, x) \right] dt$$

are solutions to the Hamilton equations.

Problem 2.10.7: Consider the problem

$$\min_{x(-1)=0, x(1)=1} \int_{-1}^1 t^2 \dot{x}(t)^2 \, dt.$$

Show that

$$x_n = \frac{1}{2} + \frac{\arctan nx}{2 \arctan n}$$

is a minimizing sequence that does not converge uniformly.

Problem 2.10.8: Let $B_1(0) \subset \mathbb{R}^2$ be a disk. Compute the Euler–Lagrange equation corresponding to the map

$$u \mapsto \int_{B_1(0)} (u_x^2 - u_y^2) \, dx dy,$$

with $u \in C(\overline{B_1}) \cap C^1(B_1)$ and $u = 0$ on $\partial B_1(0)$. Show that the solutions to the Euler–Lagrange PDE are not minimizers.

Problem 2.10.9: Let $\Omega \subset \mathbb{R}^n$ be a regular domain, and u a $C^2(\Omega) \cap C(\bar{\Omega})$ solution of $\Delta u = f$ in Ω, with $u = 0$ on $\partial\Omega$. Show that u minimize the functional $\int_\Omega \left[\frac{1}{2}\|\nabla u\|^2 + f \right] dx^1 ... dx^n$ over all C^2 functions that vanish in $\partial\Omega$.

Problem 2.10.10: Consider a rigid body with a fixed point and such that $I_1 = I_2$ (moments of inertia). Show that the kinetic energy written in the local coordinates $(\theta, \varphi, \psi) \in (0, \pi) \times (0, 2\pi) \times (0, 2\pi)$ (Euler angles) is $L = \frac{I_1}{2}(\dot\theta^2 + \dot\varphi^2 \sin^2\theta) + \frac{I_3}{2}(\dot\psi + \dot\varphi \cos\theta)^2$. Find Euler–Lagrange ODEs.

Problem 2.10.11: Let us consider two smooth functions $f : \mathbb{R}^n \to \mathbb{R}$, and $g : \mathbb{R}^n \to \mathbb{R}^m$, with $m < n$. Assume that Dg has maximal rank at all points. Let x_0 be a minimizer of $f(x)$ under the constraint $g(x) = g(x_0)$, and λ be the corresponding Lagrange multiplier. We introduce the function $F = f + \lambda g$. Show that $D^2_{x^i x^j} F(x_0) \xi^i \xi^j \geq 0$, for all vectors ξ that satisfy $D_{x^i} g(x_0) \xi^i = 0$.

Problem 2.10.12: Let us consider $\mathbb{R}^2 \setminus \{0\}$ with polar coordinates (r, θ). Show that the standard metric in $\mathbb{R}^2$ can be written in these coordinates as $g = dr^2 + r^2\dot\theta^2$. The Lagrangian of a free particle in polar coordinates is $L(r, \theta, \dot r, \dot\theta) = \frac{1}{2}(\dot r^2 + r^2\dot\theta^2)$. Determine the Euler–Lagrange ODEs and compute the corresponding Christoffel symbols.

Problem 2.10.13: Find the C^1 functions $x(t)$ for which the functional

$$I(x(\cdot)) = \int_0^T \frac{1}{x(t)} \sqrt{1 + \dot{x}^2(t)} \, dt$$

can have extrema with $x(0) = 1$ in the following cases: 1) the point $(T, x(T))$ is on the straight line $x = t - 5$; 2) the point $(T, x(T))$ is on the circle $(t - 9)^2 + x^2 = 9$.

Problem 2.10.14: The plots package contains over fifty commands that can be used to plot various Maple expressions and data structures. A brief description of some of the more useful and commonly needed commands are presented below.

```
> with(plots):
> F := plot3d(sin(x * y), x = -Pi..Pi, y = -Pi..Pi):
> G := plot3d(x + y, x = -Pi..Pi, y = -Pi..Pi):
> H := plot3d([2 * sin(t) * cos(s), 2 * cos(t) * cos(s), 2 * sin(s)], s = 0..Pi,
t = -Pi..Pi):
> display([F, G, H]);
```

3

Optimal Models Based on Energies

Motto:
"It was dark. The rain beat down far off, outside.
And my hand hurt me, like a claw
That can't be clenched.
And I forced myself to write
With my left-hand nails."
    Tudor Arghezi – *Flowers of Mildew*

In this chapter, we present some selected applications from science and engineering (brachis-
tochrone, chains, soap bubbles, elastic beam, evolutionary microstructure, particle systems,
vibrating string, vibrating membrane, Schrödinger equation), where variational calculus is
indispensable. For one-dimensional problems, the end conditions will be specified. For multi-
dimensional problems, we need boundary conditions. Traditional texts stop only at extremum
necessary conditions, because the treated problems certainly have solutions and once found
they stay the same. We also specify sufficient conditions, but we emphasize that mathemati-
cal principles are not entirely equivalent to physical ones. Details about the models based on
energies can be found in the papers [4, 6, 7, 12, 15, 16, 18, 19, 21, 22, 30, 31, 33, 34, 48, 57, 58,
71, 79, 80, 83].

3.1 Brachistochrone Problem

In 1696, Johann Bernoulli imagined the next problem of brachistochrone (in Greek,
brachistos = shortest, chrones = time).

Problem 3.1.1:   (Brachistochrone problem) Let us suppose that a material point slides
smoothly and without friction, under the action of gravity, on a thin wire, from the point
$P_0(x_0, y_0)$ to the point $P_1(x_1, y_1)$. Find out the shape of the wire, if the movement is made
in the shortest time.

Solution.   Suppose that the graph of the function $y = y(x)$ represents the shape of wire.
Let m be the mass of material point and v its velocity. At point $P(x, y)$ acts the gravitational

Variational Calculus with Engineering Applications, First Edition. Constantin Udriste and Ionel Tevy.
© 2023 John Wiley & Sons Ltd. Published 2023 by John Wiley & Sons Ltd.

force of size $F = mg$. The mechanical work is given by the formula $L = Fd$. Kinetic energy is written $T = \frac{1}{2}mv^2$. The mechanical work performed is identified with the variation of the potential energy $mg(y - y_0)$ when we move from point y_0 to point y. The corresponding variation of kinetic energy is $\frac{1}{2}m(v^2 - v_0^2)$. The theorem of mechanical work–energy requires

$$mg(y - y_0) = \frac{1}{2}m(v^2 - v_0^2).$$

Suppose that the particle of mass $m = 1$ is at rest, that is, $v_0 = 0$. Also, we know that $v = \dfrac{ds}{dt}$, where s is the curvilinear abscissa, i.e., $ds = \sqrt{1 + y'^2}\,dx$. It rests $\dfrac{ds}{dt} = \sqrt{2g(y - y_0)}$. It follows the time functional

$$T = \int_{x_1}^{x_2} \frac{ds}{v} = \frac{1}{\sqrt{2g}} \int_{x_1}^{x_2} \frac{\sqrt{1 + y'^2}}{\sqrt{y - y_0}}\,dx.$$

Since the Lagrangian $L = \dfrac{1}{\sqrt{2g}} \dfrac{\sqrt{1 + y'^2}}{\sqrt{y - y_0}}$ does not depend on the variable x, the Euler–Lagrange equation $\dfrac{\partial L}{\partial y} - \dfrac{d}{dx}\dfrac{\partial L}{\partial y'} = 0$ is replaced by the first integral $y'\dfrac{\partial L}{\partial y'} - L = k$ or

$$\frac{y'^2}{\sqrt{(y - y_0)(1 + y'^2)}} - \frac{\sqrt{1 + y'^2}}{\sqrt{y - y_0}} = k.$$

This differential equation is an equation with separate variables

$$\frac{\sqrt{y - y_0}}{\sqrt{k^2 - (y - y_0)}}\,dy = dx.$$

To get solutions we do the substitution $y = y_0 + k^2 \sin^2 \theta$. We find

$$dy = 2k^2 \sin \theta \cos \theta\,d\theta, \quad dx = 2k^2 \sin^2 \theta\,d\theta.$$

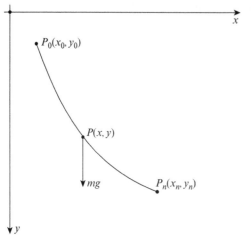

Figure 3.1 Brachistochrone.

Integrating, the Euler–Lagrange equation has the parametric solution

$$x = x_0 + \frac{1}{2}k^2(2\theta - \sin 2\theta), \quad y = y_0 + \frac{1}{2}k^2(1 - \cos 2\theta),$$

which represents a *family of cycloids*. The constant k and the interval $[0, \theta_1]$ are obtained from the condition for the cycloid to pass through the point (x_1, y_1) (see Fig. 3.1). The solution ensures the minimum of the functional since $\dfrac{\partial^2 L}{\partial y'^2} > 0$ (Lagrange-Jacobi test).

3.2 Ropes, Chains and Cables

Problem 3.2.1: **(RCC problem)** Find the shape of a flexible chain of length L that joins the points $(-\ell, h)$ and (ℓ, h), with $h > 0$, $L > 2\ell$, left free under the influence of gravity.

Solution. Physical principle: the equilibrium position of the cable corresponds to the minimum potential energy. Let ρ be the mass density per unit length, constant. The element of mass dm is localized at the point (x, y). This element of mass has the poten-tial energy $dV = (dm)gy = (\rho ds)gy$, where $ds = \sqrt{1 + y'^2}dx$. The total potential energy $V = \int_{-\ell}^{\ell} dV = \rho g \int_{-\ell}^{\ell} y\sqrt{1 + y'^2}\,dx$ is subject to an isoperimetric constraint $\int_{-\ell}^{\ell} \sqrt{1 + y'^2}\,dx = L$. In this way the function $y = y(x)$ must be a minimum point of the functional that represents the constrained potential energy. Having a problem with isoperi-metric restriction, there is a constant Lagrange multiplier λ, which changes the problem with restrictions to a problem without restrictions

$$\min_{y(\cdot)} \int_{-\ell}^{\ell} \left(\rho g y \sqrt{1 + y'^2} + \lambda \sqrt{1 + y'^2} \right) dx.$$

The new Lagrangian is

$$\mathcal{L} = (\rho g y + \lambda)\sqrt{1 + y'^2}.$$

The Euler–Lagrange equation $\dfrac{\partial \mathcal{L}}{\partial y} - \dfrac{d}{dx}\dfrac{\partial \mathcal{L}}{\partial y'} = 0$ becomes

$$\frac{d}{dx}\frac{\rho g y y' + \lambda y'}{\sqrt{1 + y'^2}} - \rho g \sqrt{1 + y'^2} = 0.$$

Since the Lagrangian $\mathcal{L}$ does not depend on the variable x, the Euler–Lagrange ODE trans-forms into the first integral $y'\dfrac{\partial \mathcal{L}}{\partial y'} - \mathcal{L} = c$ or $y'^2 = \dfrac{(\rho g y + \lambda)^2}{c^2} - 1$. With the change of function $\rho g y + \lambda = cz$, we get the equation with separate variables $\dfrac{dz}{\sqrt{z^2 - 1}} = \dfrac{\rho g}{c}\,dx$ with implicit solution $\ln\left(z + \sqrt{z^2 - 1}\right) = \dfrac{\rho g}{c}x + b$. We select $b = -\dfrac{\rho g}{c}\beta$, solve with respect to

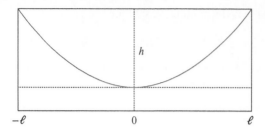

Figure 3.2 Graph of hyperbolic cosinus function.

z and we return to y. We find a solution of the form

$$y(x) = -\frac{\lambda}{\rho g} + \frac{c}{\rho g} \cosh\left(\frac{\rho g(x - \beta)}{c}\right).$$

Consequently, the shape of the chain is the graph of the hyperbolic cosinus function (see Fig. 3.2). Because in Latin a chain is called "catena", for the solution of the previous problem the name "catenary" was adopted. The constant c is fixed by the condition

$$L = \int_{-\ell}^{\ell} \sqrt{1 + y'^2}\, dx = \int_{-\ell}^{\ell} \cosh\left(\frac{\rho g x}{c}\right) dx = \frac{2c}{\rho g} \sinh\left(\frac{\rho g \ell}{c}\right).$$

The relations $y(-\ell) = y(\ell) = h$ and a suitable choice for the constant λ leads to the constant $\beta = 0$ (see Fig. 3.2). The solution found ensures the minimum of functional because the Legendre-Jacobi test, $\dfrac{\partial^2 L}{\partial y'^2} > 0$, is satisfied.

3.3 Newton's Aerodynamic Problem

The functional describing the rocket (bullet) resistance is [59]

$$J(y(\cdot)) = \int_0^R \frac{x}{1 + y'(x)^2}\, dx,$$

subject to the conditions $y =$ elevation, $x =$ section radius and

$$y(0) = L,\ y(R) = 0,\ y'(x) \leq 0,\ y''(x) \geq 0.$$

To discuss this problem we need some assumptions (see Fig. 3.3): (1) as the rocket (bullet) may rotate along its length, the nose cone must be circularly symmetric, and so we reduce the problem to one of determining the optimal profile of the nose cone; (2) the rocket's (bullet's) nose cone must have radius R at its base, and length L, and its shape should be convex; (3) air is thin, and composed of perfectly elastic particles; (4) particles will bounce off the nose cone with equal speed, and equal angle of reflection and incidence; (5) we ignore tangential friction; (6) we ignore "non-Newtonian" affects such as those from compression of the air.

These assumptions are realistic for high-altitude, supersonic flight. Newton reasoned that the profile should be a surface of revolution of a curve $x = f(y)$ about the vertical Ox axis.

Figure 3.3 Newton's bullet.

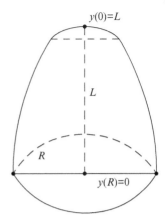

The elementary force is $F = ma$, where m means mass and a means acceleration. In our case $a = 2v \sin^2 \theta$. If we normalize, i.e., $2vm = 1$, then

$$F = \sin^2 \theta = \frac{1}{1 + \cot^2 \theta} = \frac{1}{1 + y'^2}.$$

To obtain the functional, we need to integrate over a revolution surface area at radius x, i.e., we use the area element $2\pi x dx$. Scaling to remove irrelevant constants, the functional describing the resistance is $J(y(\cdot))$ subject to the previous conditions.

The Euler–Lagrange ODE is

$$\frac{d}{dx}\left(\frac{2xy'(x)}{\left(1 + y'(x)^2\right)^2} \right) = 0,$$

or explicitly,

$$2xy'(x) = -c\left(1 + y'(x)^2\right)^2.$$

Let us find a parametric solution. Setting $y'(x) = -u(x)$, we find

$$x(u) = c\left(\frac{1}{u} + 2u + u^3\right), \quad c = const.$$

Moreover, the chain rule gives

$$\frac{dy}{du} = \frac{dy}{dx}\frac{dx}{du} = -u\frac{dx}{du} = c\left(\frac{1}{u} - 2u - 3u^3\right).$$

Finally, by integration we find

$$y(u) = c_1 - c\left(-\ln u + u^2 + \frac{3}{4}u^4\right).$$

The previous problem has a parametric solution

$$x(u) = c\left(\frac{1}{u} + 2u + u^3\right) = \frac{c}{u}\left(1 + u^2\right)^2 > 0,$$

$$y(u) = c_1 - c\left(-\ln u + u^2 + \frac{3}{4}u^4\right).$$

We add the endpoint conditions

$$y(u_1) = L, \; y(u_2) = 0; \; x(u_1) = x_1, \; x(u_2) = R,$$

but we must specify who are x_1, u_1, u_2.

Since at u_1 we must have $y(u_1) = L$, we select

$$y(u) = L - c\left(-A - \ln u + u^2 + \frac{3}{4}u^4\right).$$

Then at u_1, we need $L = L - c\left(-A - \ln u_1 + u_1^2 + \frac{3}{4}u_1^4\right)$, i.e., $A = -\ln u_1 + u_1^2 + \frac{3}{4}u_1^4$.
On the other hand, at u_2, we impose $x(u_2) = R$, $y(u_2) = 0$. It follows

$$L = c\left(-A - \ln u_2 + u_2^2 + \frac{3}{4}u_2^4\right), \; R = \frac{c}{u_2}\left(1 + u_2^2\right)^2.$$

Using $\frac{L}{R}$, we find u_2. Then u_1.

Let us plot a Newton profile. For that we select $c = 1$ and $L = 20$, divided by 10. It follows the curve (Fig. 3.4)

$$x(t) = \frac{(1+t^2)^2}{10\,t}, \; y(t) = \frac{1}{10}\left(20 + \ln t - t^2 - \frac{3}{4}t^4\right), \; 0.35 \le t \le 2.15.$$

Remark 3.1: In modern form, Newton's problem is to minimize the following simple integral functional

$$I(x(\cdot)) = \int_0^L \frac{x(t)\dot{x}(t)^3}{1 + \dot{x}(t)^2}\,dt,$$

where $x(t)$ represents the curve which generates a solid when it is rotated about the t-axis. This integral is related to the total resistance experienced by the body via the following relation

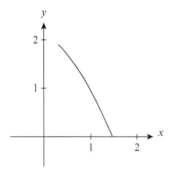

Figure 3.4 Newton profile.

$$\rho = \frac{H^2}{2} + \int_0^L \frac{x(t)\dot{x}(t)^3}{1 + \dot{x}(t)^2}\, dt.$$

At the very end of the 20th century, the problem was considered for a body whose boundary is not surface of revolution. It was shown that the removal of the hypothesis of axial symmetry allows reducing resistance: nonaxially symmetric bodies with less resistance than symmetric ones of the same length and cross-section were found.

The functional $I(x(\cdot)) = \int_0^L \frac{x(y)x'(y)^3}{1 + x'(y)^2}\, dy$ can be changed into the functional $J(y(\cdot)) = \int_0^R \frac{x}{1 + y'(x)^2}\, dx$ via the derivative of inverse function $x'(y) = \frac{1}{y'(x)}$.

Remark 3.2: Let us find an exact analytic expression for the shape of bodies exhibiting minimal resistance while moving in rarefied air surroundings. For axially symmetric convex bodies, this problem was proposed and solved by Sir Isaac Newton. If $u = u(x^1, x^2)$ is a convex function describing the shape of the body, then the resistance is the double integral functional

$$I(u) = \int_\Omega \frac{1}{1 + u_{x^1}{}^2 + u_{x^2}{}^2}\, dx^1 \wedge dx^2,$$

where Ω is the support of the function u.

3.4 Pendulums

The simplest and most instructive mechanical systems are pendulums (Figs 3.5–3.7).

3.4.1 Plane Pendulum

Let Oy be the vertical axis (Fig. 3.5). In general, the Lagrangian, the kinetic energy and the potential energy are

$$L = T - U, T = \frac{mv^2}{2}, U = mgy,$$

where m is the mass of pendulum, $v^2 = \dot{x}^2 + \dot{y}^2$ is the square of the length of the velocity, g is the acceleration of free fall. Due to the restrictions, the system is described only by the angle θ between the pendulum arm and the negative direction of the axis Oy. We observe that $v^2 = \ell^2 \dot{\theta}^2$, $x = \ell \sin \theta$, $y = -\ell \cos \theta$, where ℓ is the length of pendulum arm. Therefore, the Lagrangian in this problem becomes

$$L = m\ell^2 \left(\frac{1}{2} \dot{\theta}^2 + \omega^2 \cos \theta \right), \quad \omega^2 = \frac{g}{\ell}.$$

The Euler–Lagrange ODE is

$$\ddot{\theta} = -\omega^2 \sin \theta.$$

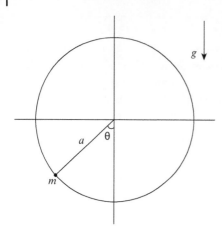

Figure 3.5 Plane pendulum.

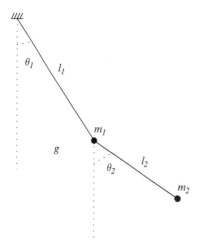

Figure 3.6 Coupled pendulums.

To solve this ODE, we need the theory of elliptical functions. When the angle is small and $\sin\theta$ is approximated by θ, the equation becomes linear, $\ddot\theta = -\omega^2\theta$ (small oscillations). The sufficient condition $\dfrac{\partial^2 L}{\partial\dot\theta^2}(\theta,\dot\theta) > 0$ is satisfied.

3.4.2 Spherical Pendulum

Let O_z axis be vertical and directed downwards. The general form of the Lagrangian is

$$L = T - U, \quad T = \frac{mv^2}{2}, \quad U = -mgz.$$

The two degrees of freedom are parameterized by the spherical angles θ and φ. The radius of the sphere is the length of the arm ℓ. Since

$$x = \ell\sin\theta\cos\varphi, \quad y = \ell\sin\theta\sin\varphi, \quad z = \ell\cos\theta,$$

we find

$$\dot x = \ell\dot\theta\cos\theta\cos\varphi - \ell\dot\varphi\sin\theta\sin\varphi$$

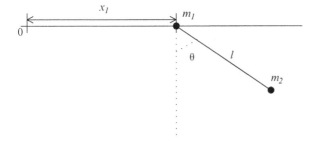

Figure 3.7 Sliding body with attached pendulum.

$$\dot{y} = \ell\dot{\theta}\cos\theta\sin\varphi + \ell\dot{\varphi}\sin\theta\cos\varphi$$

$$\dot{z} = -\ell\dot{\theta}\sin\theta.$$

It follows the square of speed

$$v^2 = \dot{x}^2 + \dot{y}^2 + \dot{z}^2 = \ell^2(\dot{\theta}^2 + \dot{\varphi}^2\sin^2\theta).$$

We deduce the Lagrangian

$$L = m\ell\left(\frac{1}{2}\dot{\theta}^2 + \frac{1}{2}\dot{\varphi}^2\sin^2\theta + \omega^2\cos\theta\right), \quad \omega^2 = \frac{g}{\ell}.$$

The two evolution Euler–Lagrange ODEs are

$$\ddot{\theta} = \dot{\varphi}^2\sin\theta\cos\theta - \omega^2\sin\theta, \quad \frac{d}{dt}(\dot{\varphi}\sin^2\theta) = 0.$$

3.4.3 Variable Length Pendulum

Suppose the length of the pendulum arm ℓ depends on time, that is, $\ell = \ell(t)$ is a given function. We start like a flat pendulum, that is

$$L = T - U, \quad T = \frac{mv^2}{2}, \quad U = mgy = -mg\ell\cos\theta.$$

Now the speed is divided into two components: the component $v_\perp = \ell\dot{\theta}$ perpendicular to the thread and the component $v_\parallel = \dot{\ell}$ parallel to the thread. It follows $v^2 = v_\perp^2 + v_\parallel^2 = \ell^2\dot{\theta}^2 + \dot{\ell}^2$. We can write the Lagrangian

$$L = m\left(\frac{1}{2}\ell^2(t)\dot{\theta}^2 + g\ell(t)\cos\theta + \frac{1}{2}\dot{\ell}^2\right).$$

The mass m is constant and the last term in Lagrangian does not change the Euler–Lagrange equation. It remains

$$\frac{d}{dt}(\ell^2(t)\dot{\theta}) = -g\ell(t)\sin\theta, \quad \ddot{\ell} = \ell\dot{\theta}^2 + g\cos\theta.$$

3.5 Soap Bubbles

Let us consider the graph of a C^2 function $y = y(x)$, fixed by the ends $y(x_1) = y_1$ and $y(x_2) = y_2$. Rotating the graph around the Ox axis we obtain a rotation surface Σ.

Problem 3.5.1: **(Soap bubbles, Fig. 3.8)** Find the curve $y = y(x)$ such that the rotation surface Σ has the minimum area.

Solution. The area element of the surface Σ is $d\sigma = 2\pi y ds = 2\pi y \sqrt{1 + y'^2}\, dx$. We obtain the area functional

$$S = \int_{x_1}^{x_2} 2\pi y(x) \sqrt{1 + y'^2(x)}\, dx$$

whose minimum we want to find out. Because the Lagrangian $L = 2\pi y \sqrt{1 + y'^2}$ does not depend on x, the Euler–Lagrange ODE $\dfrac{\partial L}{\partial y} - \dfrac{d}{dx}\dfrac{\partial L}{\partial y'} = 0$ is replaced by the first integral $y' \dfrac{\partial L}{\partial y'} - L = c$ or explicitly with

$$\frac{yy'^2}{\sqrt{1 + y'^2}} - y\sqrt{1 + y'^2} = c.$$

This differential equation is written in simplified form $\dfrac{dx}{dy} = \dfrac{c_1}{\sqrt{y^2 - c_1^2}}$, where $c_1 = -c$,

with the solution $y(x) = c_1 \cosh\left(\dfrac{x - c_2}{c_1}\right)$, where c_2 is a constant. The constants c_1 and c_2 are fixed by the conditions $y(x_1) = y_1$ and $y(x_2) = y_2$. In this way we obtain a "catenoid" (see Fig. 3.8).

The found solution ensures the minimum of functional because the condition $\dfrac{\partial^2 L}{\partial y'^2}(y(x), y'(x)) > 0$ (Legendre-Jacobi test) is satisfied.

Figure 3.8 Catenoid.

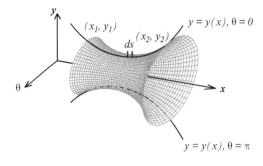

3.6 Elastic Beam

A system subject to external stresses reaches thermal equilibrium, that is, Gibbs' free energy is minimized. This thermodynamic principle allows us to find the equilibrium shape of an elastic body subject to external stresses.

As an example, we consider an elastic beam subject to lateral stress $w(x)$, as in Fig. 3.9. Ignoring the entropic contribution, Gibbs' free energy is the enthalpy H, meaning "stored elastic energy – the mechanical work done by external stress".

Assuming that the beam has the linear tension T, enthalpy can be written as a functional of a C^2 function $y(x)$, which gives the shape of the beam (accepting that $y'(x) \ll 1$), i.e.,

$$H(y(\cdot)) = \int_0^L \left(\frac{1}{2} T y'(x)^2 - w(x) y(x) \right) dx.$$

Because the equilibrium form minimizes the functional H, the function $y(x)$ is the solution of the Euler–Lagrange equation

$$T y''(x) + w(x) = 0.$$

3.7 The ODE of an Evolutionary Microstructure

The evolution of some physical systems is based on Gibbs' free energy minimization. This situation also includes the model of the phase field of a material microstructure, whose classic representative is the growth of a crystal by solidifying a melted mixture. The system can be described by a phase C^2 field $\phi(x)$, with $\phi(x) \approx 1$ if the material in point x is solid and with $\phi(x) \approx -1$ if it is liquid. The free energy of the system is the simple integral functional

$$F(\phi(\cdot)) = \frac{1}{2} \int_0^\ell \left((\phi(x) - 1)^2 (\phi(x) + 1)^2 U + \varepsilon (\phi'(x) - \alpha \phi(x))^2 \right) dx.$$

Let us consider the Lagrangian

$$L = \frac{1}{2} \left((\phi(x)^2 - 1)^2 U + \varepsilon (\phi'(x) - \alpha \phi(x))^2 \right).$$

Figure 3.9 Elastic beam.

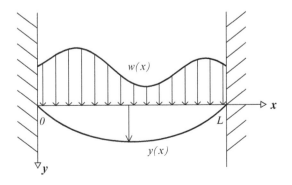

The first term counts the alternation $\phi(x) = \pm 1$, and the second term introduces a free energy penalty for the liquid-solid interface. The extremals are solutions of the Euler–Lagrange ODE

$$\frac{\partial L}{\partial \phi} - \frac{d}{dx}\frac{\partial L}{\partial \phi'} = 0.$$

3.8 The Evolution of a Multi-Particle System

Let us consider a system of N particles in $\mathbb{R}^3$, fixed by position vectors $\vec{r}_i = \vec{r}_i(t)$, $i = \overline{1,N}$ and through the masses m_i, $i = \overline{1,N}$. If $\vec{F}_i$ is the total force acting on the particle i, then the *Newton law* writes $m_i \ddot{\vec{r}}_i = \vec{F}_i$. Denote by $\delta \vec{r}_i$ the variation of the position of the particle i and using the scalar product, we can write

$$\sum_{i=1}^{N} m_i \langle \ddot{\vec{r}}_i, \delta \vec{r}_i \rangle - \sum_{i=1}^{N} \langle \vec{F}_i, \delta \vec{r}_i \rangle = 0,$$

where $\langle , \rangle$ means the scalar product. The last term represents the mechanical work $\delta W = \sum_{i=1}^{N} \langle \vec{F}_i, \delta \vec{r}_i \rangle$ of all forces $\vec{F}_i$ due to displacements $\delta \vec{r}_i$.

Remember that the unit of measure for the force $\vec{F}$ is *Newton* [N], the unit of measure for variation $\delta \vec{r}$ is *meter* [m] and the unit of measure for mechanical work W is *joule = Newton · meter*.

On the other hand we use the identity

$$\sum_{i=1}^{N} \langle m_i \ddot{\vec{r}}_i, \delta \vec{r}_i \rangle = \sum_{i=1}^{N} m_i \left(\frac{d}{dt} \langle \dot{\vec{r}}_i, \delta \vec{r}_i \rangle - \delta \left\langle \frac{1}{2}\dot{\vec{r}}_i, \dot{\vec{r}}_i \right\rangle \right).$$

If we introduce the *kinetic energy*

$$T = \sum_{i=1}^{N} \frac{1}{2} m_i \langle \dot{\vec{r}}_i, \dot{\vec{r}}_i \rangle = \sum_{i=1}^{N} \frac{1}{2} m_i v_i^2,$$

we can write

$$\delta T + \delta W = \sum_{i=1}^{N} m_i \frac{d}{dt} \langle \dot{\vec{r}}_i, \delta \vec{r}_i \rangle.$$

We add the conditions $\delta \vec{r}_i(t_1) = 0$ and $\delta \vec{r}_i(t_2) = 0$ and integrate,

$$\int_{t_1}^{t_2} (\delta T + \delta W)dt = \sum_{i=1}^{N} m_i \langle \dot{\vec{r}}_i, \delta \vec{r}_i \rangle \Big|_{t_1}^{t_2} = 0.$$

Suppose there exists a potential energy function $V = V(\vec{r}_i)$ such that $\delta W = -\delta V$. For example, the potential energy is a function of the form

$$V = \sum_{i<j} V(\|\vec{r}_i - \vec{r}_j\|).$$

Introducing the Lagrange function

$$L = T - V = \sum_{i=1}^{N} \frac{1}{2} m_i \langle \dot{\vec{r}}_i, \dot{\vec{r}}_i \rangle - V,$$

by the relation

$$\int_{t_1}^{t_2} (\delta T - \delta V) \, dt = 0,$$

we recognize the necessary condition as the integral action $I = \int_{t_1}^{t_2} L \, dt$ to be extremum (*Hamilton principle*), usually minimum. It follows the Euler–Lagrange system of equations (extremum necessary conditions)

$$\frac{\partial L}{\partial \vec{r}_i} - \frac{d}{dt} \frac{\partial L}{\partial \dot{\vec{r}}_i} = 0, \ i = \overline{1, N}.$$

Since the matrix of components $\frac{\partial^2 L}{\partial \dot{r}_i \partial \dot{r}_j} = m_i \delta_{ij}$ is positive definite, the previous condition is also sufficient for the existence of a minimum.

3.8.1 Conservation of Linear Momentum

The Lagrangian

$$L = \sum_{i=1}^{N} \frac{1}{2} m_i \langle \dot{\vec{r}}_i, \dot{\vec{r}}_i \rangle - \sum_{i<j} V(\|\vec{r}_i - \vec{r}_j\|)$$

is invariant in relation to the translation by a constant vector $\vec{\epsilon}$ of all radial vectors $\vec{r}_i$. Considering the transformation $\vec{r}_j \mapsto \vec{r}_j + \vec{\epsilon}$, with $\|\vec{\epsilon}\| \to 0$, formal we find

$$\delta L = \left\langle \sum_{j=1}^{N} \frac{\partial L}{\partial \vec{r}_j}, \vec{\epsilon} \right\rangle = \left\langle \vec{\epsilon}, \sum_{j=1}^{N} \frac{\partial L}{\partial \vec{r}_j} \right\rangle,$$

where $\langle , \rangle$ stands for scalar product. On the other hand the Lagrangian L is independent of arbitrary vector $\vec{\epsilon}$. That's why we can write

$$\left\langle \vec{\epsilon}, \sum_{j=1}^{N} \frac{\partial L}{\partial \vec{r}_j} \right\rangle = 0,$$

a relationship possible only if

$$\sum_{j=1}^{N} \frac{\partial L}{\partial \vec{r}_j} = 0.$$

From the Euler–Lagrange equations we find

$$\sum_{j=1}^{N} \frac{d}{dt} \frac{\partial L}{\partial \dot{\vec{r}}_j} = 0,$$

i.e., $\sum_{j=1}^{N} \frac{\partial L}{\partial \dot{\vec{r}}_j} = const$, which means that the linear momentum (impulse)

$$\sum_{j=1}^{N} \frac{\partial L}{\partial \dot{\vec{r}}_j} = \sum_{j=1}^{N} m_j \dot{\vec{r}}_j = P$$

conserves with respect to translations.

3.8.2 Conservation of Angular Momentum

The Lagrangian

$$L = \sum_{i=1}^{N} \frac{1}{2} m_i \left\langle \dot{\vec{r}}_i, \dot{\vec{r}}_i \right\rangle - \sum_{i<j} V(\|\vec{r}_i - \vec{r}_j\|)$$

is invariant with respect to the rotation of the system about an axis. Indeed, this Lagrangian depends only on the lengths of the velocities and the relative distance between the particles. Fix an axis and consider the rotation around it with the angle $\delta\varphi$. Since each $\vec{r}_j$ and each $\vec{v}_j = \dot{\vec{r}}_j$ is changed by a rotation in $\vec{r}_j \mapsto \vec{r}_j + \delta\vec{r}_j$, respectively $\vec{v}_j \mapsto \vec{v}_j + \delta\vec{v}_j$, we find

$$0 = \delta L = \sum_{j=1}^{N} \left(\left\langle \frac{\partial L}{\partial \vec{r}_j}, \delta\vec{r}_j \right\rangle + \left\langle \frac{\partial L}{\partial \vec{v}_j}, \delta\vec{v}_j \right\rangle \right).$$

To continue, we need $\delta\vec{r}_j$ and $\delta\vec{v}_j$. These appear from the remark that any vector $\vec{a}$, subjected to rotation about an axis of versor $\vec{e}$, with the angle $\delta\varphi$, transforms according to the rule $\vec{a} \mapsto \vec{a} + \delta\vec{a}$, where $\delta\vec{a} = \delta\varphi(\vec{e} \times \vec{a})$. Indeed, the vector $\delta\vec{a}$ must be perpendicular on $\vec{e}$ and on $\vec{a}$, with length $\|\delta\vec{a}\| = \delta\varphi \sin\theta$, where θ is the angle between $\vec{a}$ and $\vec{e}$. It rests that $\delta\vec{a} = \delta\vec{\varphi} \times \vec{a}$, where $\delta\vec{\varphi} = \delta\varphi\vec{e}$.

Returning to the above calculation, we transcribe

$$0 = \delta L = \sum_{j=1}^{N} \left(\left\langle \frac{\partial L}{\partial \vec{r}_j}, (\delta\vec{\varphi} \times \vec{r}_j) \right\rangle + \left\langle \frac{\partial L}{\partial \vec{v}_j}, (\delta\vec{\varphi} \times \vec{v}_j) \right\rangle \right)$$

or

$$0 = \left\langle \delta\vec{\varphi}, \sum_{j=1}^{N} \left(\frac{\partial L}{\partial \vec{r}_j} \times \vec{r}_j + \frac{\partial L}{\partial \vec{v}_j} \times \vec{v}_j \right) \right\rangle,$$

where $\langle,\rangle$ means the scalar product. Since the axis is arbitrary, the vector $\vec{e}$ is arbitrary and hence

$$0 = \sum_{j=1}^{N} \left(\vec{r}_j \times \frac{\partial L}{\partial \vec{r}_j} + \vec{v}_j \times \frac{\partial L}{\partial \vec{v}_j} \right).$$

Now back to the Euler–Lagrange equations and we can write

$$0 = \sum_{j=1}^{N} \left(\vec{r}_j \times \frac{d}{dt} \frac{\partial L}{\partial \vec{v}_j} + \vec{v}_j \times \frac{\partial L}{\partial \vec{v}_j} \right)$$

or

$$\frac{d}{dt} \sum_{j=1}^{N} \vec{r}_j \times \frac{\partial L}{\partial \vec{v}_j} = 0.$$

It follows the conservation of angular momentum,

$$\sum_{j=1}^{N} \vec{r}_j \times m_j \vec{v}_j = \sum_{j=1}^{N} \vec{r}_j \times \frac{\partial L}{\partial \vec{v}_j} = \text{const.}$$

3.8.3 Energy Conservation

The Lagrangian

$$L = T - V = \sum_{i=1}^{N} \frac{1}{2} m_i \langle \dot{\vec{r}}_i, \dot{\vec{r}}_i \rangle - V$$

produces the Hamiltonian (*total energy*)

$$H = T + V = \sum_{j=1}^{N} \left\langle \dot{\vec{r}}_j, \frac{\partial L}{\partial \dot{\vec{r}}_j} \right\rangle - L.$$

On the other hand, the Lagrangian does not depend on time (autonomous) and hence

$$\frac{dH}{dt} = \dots = -\frac{\partial L}{\partial t} = 0.$$

It follows $H = c$, energy conservation.

3.9 String Vibration

A string is placed under tension between two fixed points. Zero displacement is considered as the equilibrium position of the string. The string is required to move initially from the equilibrium position, then it is left free. We want the function that describes the movement of the string.

We place the string of length L on the Ox axis, with the fixed ends $x = 0$ and $x = L$. Denote by $u = u(x, t)$ the moving of the string from the current point x [m], for $x \in [0, L]$, at the time t [sec] and we assume that the function $u = u(x, t)$ is of class C^2. Then the partial derivative u_x is the *slope*, and the partial derivative u_t is the *velocity* at a current point (x, t) of the string.

We neglect the damping forces and assume that the slope satisfies $|u_x| \ll 1$. To apply Hamilton's principle we need to calculate the kinetic energy and potential energy of the string. For that, denoting by $\rho(x)[kg/m]$ the linear mass density of the string, we write the mass $\rho(x)\,dx$ of a string element of length dx. Then the kinetic energy density is $\frac{1}{2}\rho(x)u_t^2(x,t)\,dx$. Integrating over x we obtain the kinetic energy

$$T(t) = \frac{1}{2}\int_0^L \rho(x)u_t^2(x,t)\,dx, \quad \left[\frac{kg}{m}\right]\left[\frac{m}{sec}\right]^2[m] = [joule].$$

The potential energy V of the string is the sum $V = V_1 + V_2$, where V_1 is the mechanical work done to remove the string from the equilibrium position, and V_2 is the mechanical stretching work that occurs due to the fixed ends.

We assume that each element of the string is subject to a constant tension force $\tau\,[N]$. The mechanical work of deforming a string element dx, from equilibrium position, in an element $ds = \sqrt{1 + u_x^2}\,dx$, at time t, has the expression (force $\times$ distance)

$$dV_1 = \tau(ds - dx) = \tau\left(\sqrt{1 + u_x^2} - 1\right)dx.$$

Since $|u_x| \ll 1$, the approximation

$$\sqrt{1 + u_x^2} \cong 1 + \frac{1}{2}u_x^2$$

takes place and we can accept

$$dV_1 = \frac{1}{2}\tau u_x^2\,dx.$$

By integration with respect to x, it follows

$$V_1(t) = \frac{1}{2}\int_0^L \tau u_x^2\,dx, \quad [N][m] = [joule].$$

At each end the string can be seen as a linear arc, with spring constant k (k_1 at $x = 0$, and k_2 at $x = L$). Then the force of the spring is proportional to the displacement. If the displacement at one end is denoted by ξ, then the force of the spring is $k\xi$, and the elementary mechanical work is $k\xi\,d\xi$. It follows

$$V_2(t) = \int_0^{u(0,t)} k_1\xi d\xi + \int_0^{u(L,t)} k_2\xi d\xi = \frac{1}{2}k_1u^2(0,t) + \frac{1}{2}k_2u^2(L,t).$$

The Lagrangian

$$\mathcal{L}(t) = T(t) - V_1(t) - V_2(t) = \frac{1}{2}\int_0^L \left(\rho(x)u_t^2 - \tau u_x^2\right)dx - \frac{1}{2}k_1u^2(0,t) - \frac{1}{2}k_2u^2(L,t),$$

at moment t, determines the total action $I = \int_{t_1}^{t_2}\mathcal{L}(t)\,dt$, on hole interval $[t_1, t_2]$, which is transcribed as a double integral functional (minus two boundary terms = simple integrals)

$$I(u(\cdot)) = \frac{1}{2}\int_{t_1}^{t_2}\int_0^L \left(\rho(x)u_t^2(x,t) - \tau u_x^2(x,t)\right)dxdt$$

$$-\frac{1}{2}k_1 \int_{t_1}^{t_2} u^2(0,t)dt - \frac{1}{2}k_2 \int_{t_1}^{t_2} u^2(L,t)dt.$$

Problem 3.9.1: **(VS problem)** Find the C^2 critical point $u = u(x,t)$ of the total action I.

Solution. This functional (a sum between a double integral and two simple temporal integrals, at boundary) must reach a stationary point, independent of the times $t_2 > t_1$.
We use a variation of the form $U(x,t) = u(x,t) + \varepsilon h(x,t)$, where $h(x,t_1) = 0$, $h(x,t_2) = 0$, $x \in [0,L]$. Suppose that ρ and τ are constants and we introduce $\alpha^2 = \frac{\tau}{\rho}$.
Using the Lagrangian

$$\mathcal{L}_1 = \frac{1}{2}\left(\rho u_t^2(x,t) - \tau u_x^2(x,t)\right),$$

the necessary condition of extremum is the Euler–Lagrange PDE

$$\frac{\partial \mathcal{L}_1}{\partial u} - D_x \frac{\partial \mathcal{L}_1}{\partial u_x} - D_t \frac{\partial \mathcal{L}_1}{\partial u_t} = 0.$$

It follows a second-order linear PDE

$$u_{tt}(x,t) = \alpha^2 u_{xx}(x,t), \ x \in [0,L], \ t > 0,$$

known as the *wave equation*.

The VS problem is reduced to one of the following three mathematical problems:

(i) **The problem with Dirichlet boundary values**
If $h(0,t) = 0$, $h(L,t) = 0$, then initial conditions $u(x,0) = f(x)$, $u_t(x,0) = g(x)$, and two boundary conditions $u(0,t) = $ given, $u(L,t) = $ given are added to wave PDE.
(ii) **The problem with Robin boundary values**
If $h(0,t)$, $h(L,t)$ are arbitrary, then the natural boundary conditions

$$\tau u_x(0,t) - k_1 u(0,t) = 0, \ \ \tau u_x(L,t) + k_2 u(L,t) = 0$$

are added to wave PDE. The normal vector to the straight line $x = 0$ is $\vec{n} = -\vec{\tau}$, and the normal vector to the straight line $x = l$ is $\vec{n} = \vec{\tau}$. That's why the partial derivative $u_x(0,t)$ can be replaced by $-\frac{\partial u}{\partial n}(0,t)$ and $u_x(L,t)$ can be replaced by $\frac{\partial u}{\partial n}(L,t)$.
(iii) **The problem with Neumann boundary values**
In the special case $k_1 = 0, k_2 = 0$, the boundary conditions are reduced only to zero slopes at the ends, $u_x(0,t) = 0, u_x(L,t) = 0$.

Remark 3.3: The solutions of the form $u(x,t) = \phi(ax + bt)$ of wave PDE are called *solitons*. Particular case: $u(x,t) = e^{ax+bt}$.

3.10 Membrane Vibration

The equation of motion of a vibrating membrane is obtained similar to that of the vibrating string. The thin membrane extends over an Ω region that is bordered by a fixed closed curve $C = \partial\Omega$. Denote by $u = u(x, y, t)$ moving the thin elastic membrane from the equilibrium position $u = 0$. Suppose that the function $u = u(x, y, t)$ is of class C^2. Denote by $\rho(x, y)[kg/m^2]$ the mass density per unit area. Then the density of kinetic energy per unit area is $\frac{1}{2}\rho(x, y)u_t^2(x, y, t)\,[\frac{kg}{m^2}][\frac{m}{sec}]^2 = [\frac{Nm}{m^2}]$. Integrating over x and y we obtain the kinetic energy

$$T(t) = \frac{1}{2}\int_\Omega \rho(x, y)\, u_t^2(x, y, t)\, dx\, dy\ [N\,m] = [joule].$$

The potential energy V of the membrane consists of two parts, $V = V_1 + V_2$, where V_1 is the mechanical work done to remove the membrane from the equilibrium position, and V_2 is the mechanical stretching work that occurs due to the fixed edge. We assume that each area element of the membrane is subject to the constant tension force $\tau\,[N]$. The mechanical work of deforming a membrane element $dA = dxdy$, from the equilibrium position, into an element

$$d\sigma = \sqrt{1 + u_x^2 + u_y^2}\ dx\, dy,$$

at time t, has the expression (force × distance)

$$dV_1 = \tau(d\sigma - dA) = \tau\left(\sqrt{1 + u_x^2 + u_y^2} - 1\right)\, dx\, dy.$$

Suppose that $|u_x| \ll 1$, $|u_y| \ll 1$. Then the approximation

$$\sqrt{1 + u_x^2 + u_y^2} \cong 1 + \frac{1}{2}\left(u_x^2 + u_y^2\right)$$

takes place and therefore we can accept

$$dV_1 = \frac{1}{2}\tau\left(u_x^2 + u_y^2\right)\, dx\, dy.$$

By integration with respect to x and y, it follows a double integral

$$V_1(t) = \frac{1}{2}\int_\Omega \tau\left(u_x^2 + u_y^2\right)\, dx\, dy,\ [N][m] = [joule].$$

We assume that each boundary point $\partial\Omega$ is associated with a string constant $k(s)$ that varies with respect to the curvilinear abscissa s of the boundary. Accepting a linear string, the force of the string is proportional to the displacement. The total mechanical work is (simple integral)

$$V_2(t) = \frac{1}{2}\int_{\partial\Omega} k(s)u^2(s, t)\, ds,$$

where $u(s, t) = u(x(s), y(s), t)$ represents the displacement of a point on the boundary from the equilibrium position.

The Lagrangian

$$\mathcal{L}(t) = T(t) - V_1(t) - V_2(t)$$

$$= \frac{1}{2} \int_\Omega \left(\rho u_t^2 - \tau \left(u_x^2 + u_y^2 \right) \right) dxdy - \frac{1}{2} \int_{\partial\Omega} k(s) u^2(s,t) \, ds,$$

at moment t, determines the total action $I = \int_{t_1}^{t_2} \mathcal{L}(t) \, dt$, on the hole interval $[t_1, t_2]$, which is transcribed as triple integral functional (minus a boundary term = double integral)

$$I(u(\cdot)) = \frac{1}{2} \int_{t_1}^{t_2} \int_\Omega \left(\rho u_t^2 - \tau \left(u_x^2 + u_y^2 \right) \right) dxdydt$$

$$- \frac{1}{2} \int_{t_1}^{t_2} \int_{\partial\Omega} k(s) u^2(s,t) \, dsdt.$$

Problem 3.10.1: (VM problem) Find the C^2 critical point $u = u(x,y,t)$ of the total action I.

Solution. This functional (sum between a triple integral and a double integral, at boundary) must reach a stationary point, independent of the times $t_2 > t_1$.

We start with a variation of the form $U(x,y,t) = u(x,y,t) + \epsilon h(x,y,t)$, where $h(x,y,t_1) = 0$, $h(x,y,t_2) = 0$, $(x,y) \in \Omega$. Suppose that ρ and τ are constants and we introduce $\alpha^2 = \frac{\tau}{\rho}$. Using the Lagrangian

$$\mathcal{L} = \frac{1}{2} \left(\rho u_t^2 - \tau \left(u_x^2 + u_y^2 \right) \right),$$

the necessary condition of extremum is the Euler–Lagrange PDE

$$\frac{\partial \mathcal{L}}{\partial u} - D_x \frac{\partial \mathcal{L}}{\partial u_x} - D_y \frac{\partial \mathcal{L}}{\partial u_y} - D_t \frac{\partial \mathcal{L}}{\partial u_t} = 0.$$

This PDE transcribes as a second-order linear PDE

$$u_{tt}(x,y,t) = \alpha^2 \left(u_{xx}(x,y,t) + u_{yy}(x,y,t) \right), (x,y) \in \Omega, \, t > 0,$$

known as *bi-dimensional equation of waves*.

The VM problem is reduced to one of the following three mathematical problems:

(i) **The problem with Dirichlet boundary values**
If $h(x,y,t)|_{\partial\Omega} = 0$, then to the waves equation we add initial conditions $u(x,y,0) = f(x,y)$, $u_t(x,y,0) = g(x,y)$, and a boundary condition $u(x,y,t)|_{\partial\Omega} =$ given.

(ii) **The problem with elastic boundary values**

If $h(x, y, t)|_{\partial\Omega}$ is arbitrary, then to the wave equation we add a boundary condition

$$\tau\frac{\partial u}{\partial\vec{n}} + k(s)u(s, t) = 0, \ s \in \partial\Omega.$$

The normal vector $\vec{n}$ at boundary $\partial\Omega : \vec{r}(s) = x(s)\vec{\pi} + y(s)\vec{p}$ has the expression $\vec{n} = \frac{dy}{ds}\vec{\pi} - \frac{dx}{ds}\vec{p}$ (see Fig. 3.10).

(iii) **The problem with boundary free and fixed values**

In the special case $k(s) = 0$, the condition at boundary reduces to $\frac{\partial u}{\partial n} = 0$ for $(x(s), y(s)) \in \partial\Omega$. If the function $k(s)$ grows endlessly, then it is necessary that $u(s, t) = 0$ for $(x(s), y(s)) \in \partial\Omega$.

Remark 3.4: The solutions of the form $u(x, t) = \phi(ax + by + ct)$ of the vibrating membrane equation are called *solitons*. Particular case: $u(x, y, t) = e^{ax+by+ct}$.

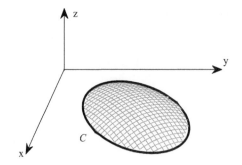

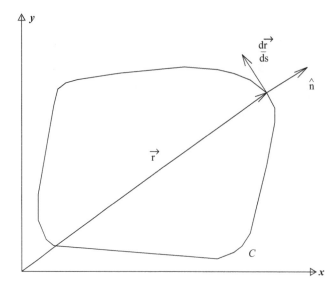

Figure 3.10 Membrane vibration.

3.11   The Schrödinger Equation in Quantum Mechanics

Let p be the classical momentum, ∇ be the gradient operator, i be the imaginary unit and $\hbar$ be the reduced Planck constant. Schrödinger constructed the moment operator using the substitution $p \mapsto -i\hbar\nabla$, and changing the Hamiltonian from classical mechanics to a Hamiltonian partial derivative operator

$$H = -\frac{\hbar^2}{2m^2}\,\nabla^2 + V(x^1, x^2, x^3),$$

which allows us to apply the Hamiltonian to the systems described by the single-time wave function $\psi(t, x^1, x^2, x^3)$.

In quantum mechanics, an electron is described by a wave function $\psi(x)$ of class C^2. For a single electron moving in a potential $V(x)$, the Hamilton operator is

$$\hat{H} = -\frac{\hbar^2}{2m}\frac{\partial^2}{\partial x^2} + V(x),$$

i.e.,

$$\hat{H}\psi(x) = -\frac{\hbar^2}{2m}\frac{\partial^2\psi}{\partial x^2}(x) + V(x)\psi(x).$$

The energy of an electron is the functional

$$E(\psi(\cdot)) = \int_{-\infty}^{\infty} \psi^*(x)\hat{H}\psi(x)\,dx, \quad \psi(-\infty) = \psi(\infty) = 0,$$

where $\psi^*(x)$ is the complex conjugate of $\psi(x)$. We add the restriction (normalization condition)

$$\int_{-\infty}^{\infty} \psi^*(x)\psi(x)\,dx = 1.$$

We use the Lagrangian

$$L = \psi^*(x)\hat{H}\psi(x) - \lambda\,\psi^*(x)\psi(x),$$

where the multiplier $\lambda = E(\psi(x)) = E$ is constant. However, the Euler–Lagrange equation is the Schrödinger equation (see eigenvalues, eigenvectors theory for differential operators)

$$\left[-\frac{\hbar^2}{2m}\frac{\partial^2}{\partial x^2} + V(x) \right]\psi(x) = E\psi(x).$$

3.11.1   Quantum Harmonic Oscillator

Let x be the displacement, ω be the angular frequency and ψ be the wave function.
The Schrödinger ODE for quantum harmonic oscillator is

$$E\psi = -\frac{\hbar^2}{2m}\frac{d^2}{dx^2}\psi + \frac{1}{2}\,m\omega^2 x^2\psi.$$

This is an example of a quantum-mechanical system whose wave function can be solved for exactly. Furthermore, it can be used to describe approximately a wide variety of other systems, including vibrating atoms, molecules, and atoms or ions in lattices, and approximating other potentials near equilibrium points. It is also the basis of perturbation methods in quantum mechanics.

In this particular case, we have a sequence of solutions

$$\psi_n(x) = \sqrt{\frac{1}{2^n n!}} \left(\frac{m\omega}{\pi\hbar}\right)^{1/4} e^{-\frac{m\omega x^2}{2\hbar}} H_n\left(x\sqrt{\frac{m\omega}{\hbar}}\right),$$

where $n \in \mathbb{N}$, and H_n are the Hermite polynomials of order n. The eigenvalues are $E_n = \left(n + \frac{1}{2}\right)\hbar\omega, \; n \in \mathbb{N}$.

3.12 Maple Application Topics

Problem 3.12.1: The Brachistochrone Lagrangian is

$$L := \frac{\sqrt{1 + diff(y(t), t)^2}}{\sqrt{y(t)}}.$$

Find Euler–Lagrange ODE.
> EulerLagrange(L, t, y(t));

Problem 3.12.2: Find the extremal of $I(y(\cdot)) = \displaystyle\int_1^2 [x^2(y'(x))^2 + y(x)]dx$, constrained by $y(1) = 1, y(2) = 1$.

Answer $y(x) = \frac{1}{2}\ln x + \frac{\ln 2}{x} + 1 - \ln 2$.

Problem 3.12.3: (**Newtonian gravity**) The Lagrangian density for Newtonian gravity is

$$L(x, t) = -\frac{1}{8\pi G}(\nabla\Phi(x, t))^2 - \rho(x, t)\Phi(x, t),$$

where Φ is the gravitational potential, ρ is the mass density and G, in $m^3 \cdot kg^{-1} \cdot s^{-2}$, is the gravitational constant. The Lagrangian L has units $J \cdot m^{-3}$.

Show that the Euler–Lagrange PDE is

$$\nabla^2\Phi(x, t)) = 4\pi G\rho(x, t),$$

i.e., Gauss's law for gravity.

Problem 3.12.4: (**Sigma Lagrangian model**) The sigma model describes the motion of a scalar point particle constrained to move on a Riemannian manifold, such as a circle or a sphere. The Lagrangian is

$$L = \frac{1}{2}g^{ij}(\phi)\,\delta^{\alpha\beta}\,\partial_\alpha\phi_i\,\partial_\beta\phi_j.$$

Find the Euler–Lagrange PDE.

Problem 3.12.5: Consider a point particle (a charged particle), interacting with the electromagnetic field. The resulting Lagrangian density for the electromagnetic field is

$$L(x,t) = -\rho(x,t)\phi(x,t) + \langle j(x,t), A(x,t)\rangle + \frac{\epsilon_0}{2} E^2(x,t) - \frac{1}{2\mu_0} B^2(x,t).$$

Find Euler–Lagrange PDEs.

Problem 3.12.6: A particle moves in the Oxy plane with a force directed towards the origin O with magnitude proportional to the distance from O. How does it move?

In Cartesian coordinates, this problem is easy. We have the Lagrangian

$$L = \frac{1}{2}((\dot{x})^2 + (\dot{y})^2) - \frac{1}{2}k(x^2 + y^2).$$

Find the Euler–Lagrange ODEs.

Problem 3.12.7: Describe the motion of two material points of mass m moving on a horizontal plane, with parallel velocities.

Solution. Let $(x^1(t), y^1(t))$ and $(x^2(t), y^2(t))$ the two points. We need to minimize the simple integral functional

$$I = \int_0^1 \left(\frac{m}{2}(\dot{x}^1(t)^2 + \dot{y}^1(t)^2) + \frac{m}{2}(\dot{x}^2(t)^2 + \dot{y}^2(t)^2) \right) dt$$

with the restriction (parallelism)

$$\dot{x}^1(t)\dot{y}^2(t) - \dot{x}^2(t)\dot{y}^1(t) = 0$$

and the boundary conditions

$$x^1(0) = 0, y^1(0) = 0, x^2(0) = 1, y^2(0) = 1$$
$$x^1(1) = -1, y^1(1) = 1, x^2(1) = 0, y^2(1) = 2.$$

We use the Lagrangian

$$L = \frac{m}{2}[(\dot{x}^1(t)^2 + \dot{y}^1(t)^2) + (\dot{x}^2(t)^2 + \dot{y}^2(t)^2)] + p(t)(\dot{x}^1(t)\dot{y}^2(t) - \dot{x}^2(t)\dot{y}^1(t)).$$

We write the Euler–Lagrange ODEs system. It follows $p(t) = c$ and the general solution (straight lines)

$$x^1(t) = \frac{mc_1 - cc_4}{m^2 - c^2} t + a_1, \quad y^1(t) = \frac{mc_2 - cc_3}{m^2 - c^2} t + b_1,$$

$$x^2(t) = \frac{mc_3 + cc_2}{m^2 - c^2} t + a_2, \quad y^2(t) = \frac{mc_4 - cc_1}{m^2 - c^2} t + b_2.$$

From the boundary conditions, we determine the constants.

Problem 3.12.8: Of all the closed plane curves of constant length ℓ, find the one that closes a domain of maximum area (hence the name "isoperimetric problem").

Solution. Let us consider a closed plane curve $\gamma : x = x(t), y = y(t), t \in [a, b], x(a) = x(b), y(a) = y(b)$, of class C^1, which borders the domain D. The area of the domain D (Green formula) is $area(D) = \dfrac{1}{2} \displaystyle\int_\gamma (x dy - y dx)$, and the length of the curve γ is $\displaystyle\int_\gamma ds$. We must maximize the functional

$$I(x(\cdot), y(\cdot)) = \frac{1}{2} \int_a^b (x(t)\dot{y}(t) - y(t)\dot{x}(t))\, dt$$

with isoperimetric constraint

$$\int_a^b \sqrt{\dot{x}(t)^2 + \dot{y}(t)^2}\, dt = \ell.$$

We use the Lagrangian

$$L = \frac{1}{2}(x(t)\dot{y}(t) - y(t)\dot{x}(t)) + p\sqrt{\dot{x}(t)^2 + \dot{y}(t)^2},$$

where p is a constant multiplier. The Euler–Lagrange ODEs system has the general solution (family of circles)

$$(x - c_1)^2 + (y - c_2)^2 = c_3^2,$$

where $c_3 > 0$. From the ends conditions, we find $a = 0, b = 2\pi$. From constant length, we obtain $p = -\dfrac{\ell}{2\pi}$.

Problem 3.12.9: (Fig. 3.11) The upper pulley is fixed in position. Both pulleys rotate freely without friction about their axles. Both pulleys are "light" in the sense that their rotational inertias are small and their rotation contributes negligibly to the kinetic energy of the system. The rims of the pulleys are rough, and the ropes do not slip on the pulleys. The gravitational acceleration is g.

Figure 3.11 Problem 3.12.9.

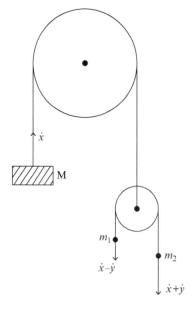

The mass M moves upwards at a rate $\dot{x}$ with respect to the upper, fixed pulley, and the smaller pulley moves downwards at the same rate. The mass m_1 moves upwards at a rate $\dot{y}$ with respect to the small pulley, and consequently its speed in laboratory space is $\dot{x} - \dot{y}$. The speed of the mass m_2 is therefore $\dot{x} + \dot{y}$ in laboratory space.

The object is to find $\ddot{x}$ and $\ddot{y}$ in terms of g.

Solution. The kinetic energy is

$$T = \frac{1}{2}M\dot{x}^2 + \frac{1}{2}m_1(\dot{x} - \dot{y})^2 + \frac{1}{2}m_2(\dot{x} + \dot{y})^2.$$

The potential energy is

$$V = g[Mx - m_1(x - y) - m_2(x + y)] + constant.$$

We write the Lagrangian $L = T - V$, and Euler–Lagrange ODEs

$$M\ddot{x} + m_1(\ddot{x} - \ddot{y}) + m_2(\ddot{x} + \ddot{y}) = -g(M - m_1 - m_2),$$
$$m_1(\ddot{x} - \ddot{y}) + m_2(\ddot{x} + \ddot{y}) = -g(m_1 - m_2).$$

These two equations can be solved to obtain $\ddot{x}$ and $\ddot{y}$.

Problem 3.12.10: **(Fig. 3.12)** A vertical torus of mass M and radius r rolls without slipping on a horizontal plane. A pearl of mass m slides smoothly around inside the torus. Describe the motion.

Solution. The vertical torus is rolling at angular speed $\dot{\phi}$. Consequently the linear speed of the center of mass of the hoop is $r\dot{\phi}$ and the pearl also shares this velocity. In addition, the pearl is sliding relative to the torus at an angular speed $\dot{\theta}$ and consequently has a component to its velocity of $r\dot{\theta}$ tangential to the torus.

The kinetic energy of the torus is $T_t = Mr^2\dot{\phi}^2$. The kinetic energy of the pearl is

$$T_p = \frac{1}{2}mr^2(\dot{\theta}^2 + \dot{\phi}^2 - 2\dot{\theta}\dot{\phi}\cos\theta).$$

Adding, we obtain the kinetic energy T of the system.

Figure 3.12 Problem 3.12.10.

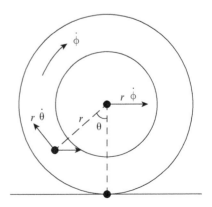

The potential energy of the torus is $V_t = 0$. The potential energy of the pearl is $V_p = -mgr \cos \theta$. It follows the Lagrangian $L = T - V$ and Euler–Lagrange ODEs

$$r(\ddot{\theta} - \ddot{\phi} \cos \theta) + g \sin \theta = 0, \quad (2M + m)\ddot{\phi} = m(\ddot{\phi} \cos \theta - \dot{\theta}^2 \sin \theta).$$

Problem 3.12.11: Let $D \subset \mathbb{R}^n$ be a bounded domain, and $f : D \to \mathbb{R}^n$ a C^∞ function with compact support. Show that the variational problem

$$\min_{u|_{\partial D}=0} I(u) = \int_D \left[\frac{1}{2}\|\nabla u\|^2 + \langle f, \nabla u \rangle + \frac{1}{1+u^2} \right] dx^1...dx^n$$

admits a minimizer in $W_0^{1,2}(D)$ (Sobolev space).

Problem 3.12.12: Determine the Lagrangian of point particle constrained to the cone $z^2 = x^2 + y^2$.

Problem 3.12.13: The classical potential $V(x) = \frac{k}{2}x^2$ for a harmonic oscillator is derivable from Hooke's law. Let ω be the natural frequency, k be the spring constant and m be the mass of the body. They are related by $\omega = \sqrt{\frac{k}{m}}$. Write the Schrödinger equation and look for solutions.

Solution. The potential for the harmonic oscillator is $V(x) = \frac{1}{2}m\omega^2 x^2$. Placing this potential in the one-dimensional, time-independent Schrödinger equation, we obtain

$$-\frac{\hbar}{2m}\frac{d^2\psi}{dx^2}(x) + V(x)\psi(x) = E\psi(x),$$

or

$$\frac{d^2\psi}{dx^2}(x) + \left(\frac{2mE}{\hbar^2} - \frac{m^2\omega^2}{\hbar^2}x^2 \right)\psi = 0.$$

This Equation for the Quantum Harmonic Oscillator is a second-order differential equation that can be solved using a power series.

Problem 3.12.14: (Potential energy of a particle) A particle with mass m and with zero energy has a time-independent wave function $\psi(x) = axe^{-\frac{x^2}{L^2}}$, where A and L are constants. Determine the potential energy $U(x)$ of the particle.

Solution. We use the time-independent Schrödinger equation for the wave function $\psi(x)$ of a particle of mass m in a potential $U(x)$:

$$-\frac{\hbar}{2m}\frac{d^2\psi}{dx^2}(x) + U(x)\psi(x) = E\psi(x).$$

By substitution, we obtain $U(x) = \frac{2\hbar}{mL^4}\left(x^2 - \frac{3L^2}{2} \right)$. Geometrically, $U(x)$ is a parabola centered at $x = 0$ with $U(0) = -\frac{3\hbar^2}{mL^2}$.

4

Variational Integrators

Motto:
"Never yet had autumn so enticing shone
To our soul – to eager to receive death's blade.
With pale silken fabrics all the plain's been sewn.
While for clouds the tree-leaves weave some new brocade."
Tudor Arghezi – *Never Yet Had Autumn...*

In the mathematical field of numerical methods in ODEs or PDEs, a variational integrator is a method that refers to discrete Euler–Lagrange equations or to discrete Hamilton equations. This explains time integration of Lagrangian dynamical systems or of Hamiltonian dynamical systems – an important computational tool at the core of most physics-based animation techniques. Several features make this particular time integrator highly desirable for computer animation: it numerically preserves important invariants, such as linear and angular momenta; the symplectic nature of the integrator also guarantees a correct energy behavior, even when dissipation and external forces are added. Some details about variational integrators can be found in the papers [4, 9, 19, 25, 29, 41, 43, 44, 47, 49, 55, 59, 60, 62, 69, 70, 71, 76, 77, 81, 82].

ODEs and PDEs that arise from realistic models of the natural world are generally nonlinear. In this chapter, we discuss also the linearization of a nonlinear ODE or PDE about a known solution. The solution to the linear problem does not solve the original problem, it only approximately solves it, and the linear approximation solution is best near the known solution of nonlinear equation (see [1, 74]).

4.1 Discrete Single-time Lagrangian Dynamics

Functionals may have many geometric or structural features. Integrators must balance costs, local, global and long-time errors, stability and structural preservation.

Denote $x = (x^i) \in \mathbb{R}^n$. Let $\gamma = \{(t, x) \mid x = x(t), t_0 \le t \le t_1\}$ be a curve in $\mathbb{R} \times \mathbb{R}^n$ (graph of the curve $x = x(t)$ in $\mathbb{R}^n$). Suppose Γ is the set of all curves γ which join the fixed points (t_0, x_0) and (t_1, x_1).

Variational Calculus with Engineering Applications, First Edition. Constantin Udriste and Ionel Tevy.
© 2023 John Wiley & Sons Ltd. Published 2023 by John Wiley & Sons Ltd.

A C^2 function

$$L : \mathbb{R} \times \mathbb{R}^n \times \mathbb{R}^n \to \mathbb{R}, \ (t, x, \dot{x}) \mapsto L(t, x, \dot{x})$$

is called *Lagrangian density energy*.

Definition 4.1: A C^2 curve γ is called extremal of the functional

$$E(\gamma) = \int_{t_0}^{t_1} L(t, x(t), \dot{x}(t)) \, dt, \ \gamma \in \Gamma$$

if $x = x(t)$ is a solution of Euler–Lagrange ODEs

$$\frac{\partial L}{\partial x^i} - \frac{d}{dt}\frac{\partial L}{\partial \dot{x}^i} = 0, \ i = \overline{1, n}.$$

This is in fact a differential system of n ODEs of at most second order in the unknown $x(t)$. When the ODEs system is of second order, the solutions depend on $2n$ arbitrary constants. For fixing these constants we can use the $2n$ boundary conditions $x(t_0) = x_0$, $x(t_1) = x_1$.

The Euler–Lagrange ODEs do not depend on the choice of the system of coordinates (tensorial character).

A regular Lagrangian L is in Legendrian duality with the Hamiltonian (total energy)

$$H(t, x, \dot{x}) = \dot{x}^i \frac{\partial L}{\partial \dot{x}^i}(t, x, \dot{x}) - L(t, x, \dot{x}).$$

The dual variable to velocity $\dot{x}(t)$ is the momentum $p_i(t) = \frac{\partial L}{\partial \dot{x}^i}(t, x, \dot{x})$. Of course, here we used Einstein convention of summation with respect to the index i (no index appears more than twice in the equation; the index which is summed over is called *dummy index*). A single-time Hamiltonian H is conserved along the extremals only if it does not depend on the parameter t.

Remark 4.1: The method of solving an Euler–Lagrange ODE system by discretization (as example, via the Runge–Kutta method) is not appropriate because we accumulate too many errors, and the numerical solution is far from the exact solution. Exceptions are cases in which the variational integrator coincides with the discretization of the Euler–Lagrange equation. These ideas are closely related to the fact that numerical integration can be done with errors no matter how small, while numerical differentiation sometimes produces large errors. More subtly, numerical integration switches to uniform convergence, while numerical derivation does not switch to uniform convergence.

The way to go is to discretize the action, write the discrete Euler–Lagrange equations (variational integrator) and then numerically solve them [25, 81]. Older books about "Finite Difference Methods for Ordinary Differential Equations" could not take this remark into account and have discretized any type of ODE, not taking into account that some come from variational principles.

The previous remarks can be summed up by the following non-commutating diagrams

$$Lagrangian \xrightarrow{\ discretization\ } discrete\ Lagrangian$$
$$\downarrow$$
$$variational\ integrator,$$

$$Lagrangian$$
$$\downarrow$$
$$E\text{-}L\ ODEs \xrightarrow[\ discretization\ ]{} discrete\ E\text{-}L\ ODEs.$$

For Lagrangians that are linear in the velocities, the continuous Euler–Lagrange ODE is of first order, but the variational integrator involves three points, i.e., it is of second order. This means that we are dealing with two-step methods; the difference equation needs one more point of initial data than the ODE.

The discretization of the Lagrangian L (discrete version of the integral $E(\gamma)$) can be made by using the midpoint rule which consists of the substitution of the time t by kh, of the point x with $\dfrac{x_{k+1} + x_k}{2}$ and the velocity $\dot{x}$ by $\dfrac{x_{k+1} - x_k}{h}$, where h is the time step.

One obtains the discrete Lagrangian

$$L_d : \mathbb{R} \times \mathbb{R}^n \times \mathbb{R}^n \to \mathbb{R}, \ L_d(kh, u, v) = L\left(kh, \frac{u+v}{2}, \frac{v-u}{h}\right).$$

Let us denote

$$L_d(k) = L\left(kh, \frac{x_{k+1} + x_k}{2}, \frac{x_{k+1} - x_k}{h}\right).$$

This determines a discrete action (discrete simple integral = sum)

$$S : \mathbb{R} \times (\mathbb{R}^n)^{N+1} \to \mathbb{R}, \ S(h; x_0, x_1, ..., x_N) = \sum_{k=0}^{N-1} L_d(k)\, h,$$

with $x_k \in \mathbb{R}^n$. The discrete variational principle characterizes the sequence of points

$$(x_0, x_1, ..., x_N) \in (\mathbb{R}^n)^{N+1}$$

for which the action S is stationary, for any family $x_k(\varepsilon) \in \mathbb{R}^n, \varepsilon \in I \subset \mathbb{R}, 0 \in I$, with $x_k(0) = x_k$, and x_0, x_N fixed points. Using the variation of first order, we obtain the discrete vectorial equations

$$\frac{1}{2}\frac{\partial}{\partial x}(L_d(k) + L_d(k-1)) - \frac{1}{h}\frac{\partial}{\partial \dot{x}}(L_d(k) - L_d(k-1)) = 0, \quad k = \overline{1, N-1}.$$

This equation seems to be a formal discretization of the Euler–Lagrange vector equation

$$\frac{\partial L}{\partial x} - \frac{d}{dt}\frac{\partial L}{\partial \dot{x}} = 0,$$

considering $\dfrac{\partial L}{\partial x}$ and $\dfrac{\partial L}{\partial \dot{x}}$ as independent variables, and using the above L_d function.

To solve the previous discrete vectorial equation, we have

Theorem 4.1: *The finite sequence* (x_k), $k = \overline{0, N}$, *is stationary for the action S if and only if it is generated by the discrete Euler–Lagrange equations (variational integrator)*

$$D_1 L_d(kh, x_k, x_{k+1}) + D_2 L_d(kh, x_{k-1}, x_k) = 0,$$

where

$$D_1 L_d = \frac{\partial L_d}{\partial x_k^i}(kh, x_k, x_{k+1}), \quad D_2 L_d = \frac{\partial L_d}{\partial x_k^i}(kh, x_{k-1}, x_k),$$

$i = \overline{1, n}; \ k = \overline{1, N-1}.$

Here we used the notation $D_j f$ to refer to the partial derivative of f with respect to its j-th x argument.

The transfer rule inside variational integrator is

$$(x_{k-1}, x_k) \mapsto (x_k, x_{k+1}).$$

To solve the discrete Euler–Lagrange equations, we denote

$$A_i(k) = \frac{\partial L_d}{\partial x_k^i}(kh, x_{k-1}, x_k), \quad f_i(u) = \frac{\partial L_d}{\partial x_k^i}(kh, x_k, u) + A_i(k),$$

$i = \overline{1, n}; \ u = (u^1, ..., u^n); \ F = (f_1, ..., f_n)$. Then the discrete Euler–Lagrange system transfers into a nonlinear equation system $F(u) = 0$ at each step k. The solution of this system can be approximated by using $u(1) = x_k$ in the Newton method

$$J_F(u(e)) \begin{pmatrix} u^1(e+1) \\ u^2(e+1) \\ ... \\ u^n(e+1) \end{pmatrix} = J_F(u(e)) \begin{pmatrix} u^1(e) \\ u^2(e) \\ ... \\ u^n(e) \end{pmatrix} - \begin{pmatrix} f_1(u(e)) \\ f_2(u(e)) \\ ... \\ f_n(u(e)) \end{pmatrix},$$

for $e = 1, 2, ..., \bar{e}$. Here J_F means the Jacobi matrix of the function F.

The discrete Lagrangian L_d produces the discrete Hamiltonian

$$H_d(k) = \frac{x_k^i - x_{k-1}^i}{h} \frac{\partial L_d}{\partial \dot{x}^i}(kh, x_{k-1}, x_k) - L_d(kh, x_{k-1}, x_k)$$

(sum over the index i).

Remark 4.2: In essence, variational integrator (the Euler's method of difference equation) transforms an integral into a finite sum and then the functional problem into an extremum problem for a function of several real variables. Thus we come to extrema problems for sums of the form

$$S(x_0, x_1, ..., x_N) = \sum_{k=0}^{N-1} L_d(k, k+1),$$

which come or not from continuous Lagrangians.

Example 4.1: (**Fibonacci sequence**) *This sequence is obtained finding the extremum point for action function*

$$S(x_0, x_1, \ldots, x_N) = \sum_{k=0}^{N-1} \frac{(-1)^k}{2} ((x_{k+1} - x_k)^2 + x_k^2).$$

Let us consider the discrete Lagrangian

$$L_d(x_{k-1}, x_k) = \frac{(-1)^{k-1}}{2} ((x_k - x_{k-1})^2 + x_{k-1}^2),$$

and via $k \mapsto k + 1$,

$$L_d(x_k, x_{k+1}) = \frac{(-1)^k}{2} ((x_{k+1} - x_k)^2 + x_k^2).$$

We compute the derivatives

$$D_2 L_d(x_{k-1}, x_k) = \frac{\partial L_d}{\partial x_k} = (-1)^{k-1}(x_k - x_{k-1}),$$

$$D_1 L_d(x_k, x_{k+1}) = \frac{(-1)^k}{2} [-2(x_{k+1} - x_k) + 2x_k].$$

By addition, it follows the variational integrator (Fibonacci sequence)

$$x_{k+1} = x_k + x_{k-1}.$$

> rsolve(x(0) = 0, x(1) = 1, x(n + 1) = x(n) + x(n − 1), x);

$$x(n) = \frac{\sqrt{5}}{5} \left(\frac{1 + \sqrt{5}}{2} \right)^n - \frac{\sqrt{5}}{5} \left(\frac{1 - \sqrt{5}}{2} \right)^n.$$

The generating function for Fibonacci numbers is

$$X(z) = \sum_0^\infty x(n) z^n = \ldots = \frac{1}{1 - z - z^2}.$$

Problem 4.1.1: (**Open problem**) Does the Fibonacci sequence discrete Lagrangian

$$L_d(x_k, x_{k+1}) = \frac{(-1)^k}{2} ((x_{k+1} - x_k)^2 + x_k^2)$$

is intrinsically discrete since it does not originate in the discretization of a continuous Lagrangian?

Problem 4.1.2: **(Queue sequence)** This sequence is derived as variational integrator for

$$S(x_0, x_1, \dots, x_N) = \sum_{k=0}^{N-1} \frac{1}{2} \rho^{k+2} (x_{k+1} - x_k)^2,$$

so is to consider the discrete Lagrangian

$$L_d(x_k, x_{k+1}) = \frac{1}{2} \rho^{k+2} (x_{k+1} - x_k)^2.$$

We compute

$$D_1 L_d(x_k, x_{k+1}) = -\rho^{k+2} (x_{k+1} - x_k).$$

On the other hand

$$L_d(x_{k-1}, x_k) = \frac{1}{2} \rho^{k+1} (x_k - x_{k-1})^2,$$

and

$$D_2 L_d(x_{k-1}, x_k) = \frac{1}{2} \rho^{k+1} (x_k - x_{k-1}).$$

Denoting $\rho = \frac{\lambda}{\mu}, \lambda > 0, \mu > 0$ and adding, we obtain the integrator

$$\lambda x_{k+1} - (\lambda + \mu) x_k + \mu x_{k-1} = 0$$

(for details see [69]). Suppose $\lambda = 1, \mu = 1$. Then we can use Maple command to produce the arithmetic mean:

> rsolve($x(n + 1) - 2 * x(n) + x(n - 1) = 0, x(n)$);

The arithmetic mean is often used to estimate future performances.

Suppose we have a solution of the form $x_k = r^k$. It follows the characteristic equation $\lambda r^2 - (\lambda + \mu)r + \mu = 0$, with the roots $r_1 = 1$ and $r_2 = \frac{\mu}{\lambda}$, for $\lambda \neq 0$. Hence, one obtains $x_k = c_1 + c_2 r_2^k$.

4.2 Discrete Hamilton's Equations

One of the chief virtues of the Lagrangian equations of motion is that they have tensorial character under an arbitrary point transformation. Hamilton's equations of motion not only share this virtue but they take it to a higher level: they have tensorial character under certain more general transformations $(x, p) \mapsto (f(x, p), g(x, p))$. Although we derived the Hamilton equations from the Euler–Lagrange equations, the Hamiltonian formalism can be viewed as independent of the Lagrangian formalism.

The Hamiltonian formulation of mechanics is in many ways more powerful than the Lagrangian formulation. Among the advantages of Hamiltonian mechanics we note that: it leads to powerful geometric techniques for studying the properties of dynamical systems; it allows a much wider class of coordinates than either the Lagrange or Newtonian formulations; it allows for the most elegant expression of the relation between symmetries and conservation laws; it leads to many structures that can be viewed as the macroscopic ("classical") imprint of quantum mechanics. Although the Hamiltonian form of mechanics is logically independent of the Lagrangian formulation, traditionally it is convenient and instructive to introduce the Hamiltonian formalism via transition from the Lagrangian formalism. The most basic change we encounter when passing from Lagrangian to Hamiltonian methods is that the "arena" we use to describe the equations of motion is no longer the configuration space, but rather the momentum phase space.

We underline that the Hamilton's equations are immune to discretization: whether we discretize the Hamilton action and make the discrete Hamilton equations or we make the Hamilton equations and discretize them is one and the same thing. The scientific world does not comment on this statement and goes straight to symplectic integration of Hamiltonian systems (for details, see [25]).

The Euler–Lagrange (at most of second order) and Hamilton equations (always of first order) are equivalent only when the Legendre transform is a global diffeomorphism (that is, the system is hyper-regular). The relationship between Lagrangian and Hamiltonian dynamics is of particular importance when the system is not hyper-regular (that is, when it is degenerated) and so the two theories are not entirely equivalent.

The Legendre transform

$$L(t, x(t), \dot{x}(t)) = \dot{x}^i(t)\, p_i(t) - H(t, x(t), p(t))$$

changes the initial simple integral functional

$$I(x(\cdot)) = \int_{t_0}^{t_1} L(t, x(t), \dot{x}(t))\, dt$$

in a new functional

$$J(x(\cdot), p(\cdot)) = \int_{t_0}^{t_1} [\dot{x}^i(t)\, p_i(t) - H(t, x(t), p(t))]\, dt$$

and the Euler–Lagrange equations of the new functional $J(x(\cdot), p(\cdot))$ are even the Hamilton ODEs

$$\dot{x}^i(t) = \frac{\partial H}{\partial p_i}(t, x(t), p(t)), \quad \dot{p}_i(t) = -\frac{\partial H}{\partial x^i}(t, x(t), p(t)), \quad i = \overline{1, n}.$$

Theorem 4.2: *The discrete Hamilton equations attached to the discretized action coming from $J(x(\cdot), p(\cdot))$ coincide with the central discretized Hamilton equations.*

Proof. The discrete version of the Lagrangian

$$\mathcal{L} = p_i(t)\dot{x}^i(t) - H(t, x(t), p(t))$$

is

$$\mathcal{L}_d(kh, x_k, x_{k+1}, p_k, p_{k+1}) = \frac{p_{ik+1} + p_{ik}}{2} \frac{x_{k+1}^i - x_k^i}{h} - H_d(kh, x_k, x_{k+1}, p_k, p_{k+1})$$

(sum with respect to the index i). Let us denote

$$G_d(k) = \left\langle \frac{p_{k+1} + p_k}{2}, \frac{x_{k+1} - x_k}{h} \right\rangle - H(k),$$

where

$$H(k) = H\left(kh, \frac{x_{k+1} + x_k}{2}, \frac{p_{k+1} + p_k}{2}\right).$$

This produces the discrete action

$$S(h; x_0, x_1, ..., x_N; p_0, p_1, ..., p_N) = \sum_{k=0}^{N-1} G_d(k)\, h.$$

Here $x_k \in \mathbb{R}^n$, $p_k \in \mathbb{R}^n$. The discrete variational principle characterizes the finite sequence of points

$$(x_0, x_1, ..., x_N) \in (\mathbb{R}^n)^{N+1}, \quad (p_0, p_1, ..., p_N) \in (\mathbb{R}^n)^{N+1}$$

for which the action S is stationary, for any family $x_k(\varepsilon) \in \mathbb{R}^n$, $p_k(\varepsilon) \in \mathbb{R}^n$, $\varepsilon \in I \subset \mathbb{R}$, $0 \in I$, with $x_k(0) = x_k$, and x_0, x_N fixed points, and we still have $p_k(0) = p_k$, and p_0, p_N fixed points. Using the variation of first order, the finite sequences (x_k), and (p_k), $k = \overline{0, N}$, are stationary for the action S if and only if they are generated by the discrete vectorial equations (central difference approximation)

$$\frac{x_{k+1} - x_{k-1}}{2h} = \frac{1}{2}\frac{\partial}{\partial p}(H(k) + H(k-1)),$$

$$\frac{p_{k+1} - p_{k-1}}{2h} = -\frac{1}{2}\frac{\partial}{\partial x}(H(k) + H(k-1)),$$

for $k = \overline{1, N-1}$. We remark that these equations are the same with the discrete Hamilton equations obtained via central difference approximation. The central differences are significantly better. In other words it is the best to use central differences whenever possible. □

Problem 4.2.1:   Show that the so-called symplectic Euler method

$$p_{k+1} = p_k - h\frac{\partial H}{\partial x}(p_{k+1}, x_k), \quad x_{k+1} = x_k + h\frac{\partial H}{\partial p}(p_{k+1}, x_k)$$

is a symplectic method of order 1.

Solution.   Differentiation with respect to (p_k, x_k) produces

$$\begin{pmatrix} I + hH_{xp}^T & 0 \\ -hH_{pp} & I \end{pmatrix}\begin{pmatrix} \dfrac{\partial(p_{k+1}, x_{k+1})}{\partial(p_k, x_k)} \end{pmatrix} = \begin{pmatrix} I & -hH_{xx} \\ 0 & I + hH_{xp} \end{pmatrix},$$

where the matrices $H_{xp}, H_{pp}, ...$ are evaluated at (p_k, x_k). Using the matrices

$$A = \begin{pmatrix} \dfrac{\partial(p_{k+1}, x_{k+1})}{\partial(p_k, x_k)} \end{pmatrix}, \quad J = \begin{pmatrix} 0 & I \\ -I & 0 \end{pmatrix},$$

we check the symplecticity condition $A^T J A = J$.

Remark 4.3: Recall that the Euler–Lagrange equations are computed from partial derivatives of the Lagrangian, which appear in the differential

$$dL = \frac{\partial L}{\partial x^i} dx^i + \frac{\partial L}{\partial \dot{x}^i} d\dot{x}^i + \frac{\partial L}{\partial t} dt.$$

The key feature of the Legendre transformation $L \mapsto H$ is that the differentials $d\dot{x}^i$ have dropped out of dH:

$$dH = \dot{x}^i dp_i + p_i d\dot{x}^i - \frac{\partial H}{\partial x^i} dx^i - \frac{\partial H}{\partial \dot{x}^i} d\dot{x}^i + \frac{\partial H}{\partial t} dt$$

$$= \dot{x}^i dp_i - \frac{\partial H}{\partial x^i} dx^i + \frac{\partial H}{\partial t} dt.$$

4.3 Numeric Newton's Aerodynamic Problem

The Newton functional

$$J(x(\cdot)) = \int_0^R \frac{t}{1 + \dot{x}(t)^2} dt,$$

is subject to the mixed conditions

$$x(0) = L, \ x(R) = 0, \ \dot{x}(t) \leq 0, \ \ddot{x}(t) \geq 0.$$

Since the Lagrangian is $L = \frac{t}{1+\dot{x}(t)^2}$, we obtain easily the discretized Lagrangian

$$L_d = \frac{kh/2}{1 + \left(\frac{x_{k+1}-x_k}{h}\right)^2}, \ t_k = kh.$$

We use the general form of the variational integrator

$$\frac{\partial L_d}{\partial x_k}(kh, x_{k-1}, x_k) + \frac{\partial L_d}{\partial x_k}(kh, x_k, x_{k+1}) = 0.$$

Replacing

$$\frac{\partial L_d}{\partial x_k} = \frac{k}{\left(1 + \left(\frac{x_{k+1}-x_k}{h}\right)^2\right)^2} \frac{x_{k+1} - x_k}{h},$$

we find

$$\frac{(k-1)(x_k - x_{k-1})}{\left(1 + \left(\frac{x_k-x_{k-1}}{h}\right)^2\right)^2} - \frac{k(x_{k+1} - x_k)}{\left(1 + \left(\frac{x_{k+1}-x_k}{h}\right)^2\right)^2} = 0.$$

Denoting

$$F(k-1) = \frac{(k-1)(x_k - x_{k-1})}{\left(1 + \left(\frac{x_k-x_{k-1}}{h}\right)^2\right)^2},$$

the recurrence is reduced to $F(k-1) - F(k) = 0$, and hence $F(k) = F(1)$.

To solve the recurrence relation for the specified function, we use the Maple commands:

```
> rsolve(F(n − 1) − F(n) = 0, F(n));
F(n) = F(1).
> h := .1;
```

$$> \text{rsolve}\left((n-1)(f(n) - f(n-1))/\left(1 + \left(\frac{f(n)-f(n-1)}{h}\right)^2\right)^2 = F(1)\right)$$

4.4 Discrete Multi-time Lagrangian Dynamics

The multi-time calculus of variation is of course not new. Let $\Omega \subset \mathbb{R}^m$ be a relatively compact domain. Let

$$\varphi = \{(t,x) \mid x = x(t), t \in \Omega, x \in \mathbb{R}^n\}$$

be a parameterized sheet in the product space $\mathbb{R}^m \times \mathbb{R}^n$. Denote by Φ the set of all parameterized sheets φ satisfying the boundary condition $\varphi|_{\partial\Omega} = f$, where f is a given function.

Introduce the Jacobian matrix $x_t = \frac{\partial x}{\partial t}$. By components we write $x_{t^\alpha}^i = \frac{\partial x^i}{\partial t^\alpha}$. A C^2 function

$$L : \mathbb{R}^m \times \mathbb{R}^n \times \mathbb{R}^{mn} \to \mathbb{R}, \quad (t,x,x_t) \mapsto L(t,x,x_t)$$

is called *Lagrangian density energy*.

Definition 4.2:   A C^2 parameterized sheet φ is called extremal of the multiple integral functional

$$E(\varphi) = \int_\Omega L(t,x,x_t)\, dt^1 \wedge \dots \wedge dt^m$$

if φ is a solution of Euler–Lagrange equations

$$\frac{\partial L}{\partial x^i} - \frac{\partial}{\partial t^\alpha}\frac{\partial L}{\partial x_\alpha^i} = 0, \quad i = \overline{1,n}, \ \alpha = \overline{1,m}$$

(Einstein summation with respect to the index α).

This is a PDEs system with n partial differential equations, each of at most second order. If all are of the second order, then the solutions depend on $2n$ arbitrary functions. Suppose Ω is a hyper-rectangle fixed by the opposite diagonal points t_0, t_1 in $\mathbb{R}^m$. For fixing one solution we use boundary conditions either of the form $x(t)|_{\partial\Omega} = f$ or, in some special cases, of the form $x(t_0) = x_0, x(t_1) = x_1$.

The Euler–Lagrange PDEs do not depend on the choice of the system of coordinates (tensorial character of Euler–Lagrange operator).

Remark 4.4: The method of solving an Euler–Lagrange PDE system by discretization is not appropriate because we accumulate too many errors, and the numerical solution is far from the exact solution. Exceptions are cases in which the variational integrator coincides with the discretization of the Euler–Lagrange PDE. These ideas are closely related to the fact that numerical integration can be done with errors no matter how small, while numerical differentiation sometimes produces large errors. Furthermore, numerical integration switches to uniform convergence, while numerical derivation does not switch generally to uniform convergence.

The way to go is to discretize the action, write the discrete Euler–Lagrange equations (variational integrator) and then numerically solve them [25, 81]. Older books about "Finite Difference Methods for Partial Differential Equations" could not take this remark into account and have discretized any type of PDE, not taking into account that some come from variational principles.

The previous remarks can be summed up by the following non-commutating diagrams

$$
\begin{array}{ccc}
Lagrangian & \xrightarrow{\ discretization\ } & discrete\ Lagrangian \\
& & \downarrow \\
& & variational\ integrator,
\end{array}
$$

$$
\begin{array}{ccc}
Lagrangian & & \\
\downarrow & & \\
E\text{–}L\ PDEs & \xrightarrow[\ discretization\ ]{} & discrete\ E\text{–}L\ PDEs.
\end{array}
$$

For Lagrangians that are linear in the partial velocities, the continuous Euler–Lagrange PDE is of first order, but the variational integrator involves three points, i.e., it is of second order. This means that we are dealing with two-step methods; the difference equation needs one more point of initial data than the PDE.

A regular Lagrangian L is in duality with the Hamiltonian (Einstein summation with respect to j and β)

$$
H(t^\alpha, x^i, x_\alpha^i) = x_\beta^j \, \frac{\partial L}{\partial x_\beta^j}(t^\alpha, x^i, x_\alpha^i) - L(t^\alpha, x^i, x_\alpha^i).
$$

The dual variable to partial velocity x_α^i is the matrix momentum $p_i^\alpha(t) = \frac{\partial L}{\partial x_\alpha^i}(t^\beta, x^i, x_\beta^i)$. A multi-time Hamiltonian H is not conserved along the extremals.

To simplify, suppose $m = 2$. The discretization of the Lagrangian L can be made by using the centroid rule which consists in the substitution of the point $t = (t^1, t^2)$ with $(kh_1, \ell h_2)$, of the point x with arithmetic mean $\frac{1}{3}(x_{k\ell} + x_{k+1\ell} + x_{k\ell+1})$, and of partial velocities x_α, $\alpha = 1, 2$, by $\frac{1}{h_1}(x_{k+1\ell} - x_{k\ell})$ respectively $\frac{1}{h_2}(x_{k\ell+1} - x_{k\ell})$. One obtains the discrete Lagrangian

$$
L_d : \mathbb{R}^2 \times (\mathbb{R})^3 \to \mathbb{R}, \ L_d(u, v, w) = L\left(kh_1, \ell h_2, \frac{u+v+w}{3}, \frac{v-u}{h_1}, \frac{w-u}{h_2}\right).
$$

This determines the two-dimensional discrete action (discretized double integral = double sum)

$$S(h_1, h_2, A) = \sum_{k=0}^{M-1} \sum_{\ell=0}^{N-1} L(kh_1, \ell h_2; x_{k\ell}, x_{k+1\ell}, x_{k\ell+1}),$$

where $x_{k\ell} \in (\mathbb{R})^{(M+1)(N+1)}$ and

$$A = \begin{pmatrix} x_{00} & x_{01} & \cdots & x_{0N} \\ x_{10} & x_{11} & \cdots & x_{1N} \\ \cdots & \cdots & \cdots & \cdots \\ x_{M0} & x_{M2} & \cdots & x_{MN} \end{pmatrix}.$$

We fix the 2-step (h_1, h_2) and suppose that the Lagrangian is autonomous. The discrete variational principle consists in finding the matrix A for which the action S is stationary, for any family $x_{k\ell}(\varepsilon) \in \mathbb{R}^n, k = \overline{1, M-1}; \ell = \overline{1, N-1}, \varepsilon \in I \subset \mathbb{R}, 0 \in I$ with $x_{k\ell}(0) = x_{k\ell}$, and fixed lines

$$(x_{00}, x_{01}, ..., x_{0N}), \quad (x_{M0}, x_{M1}, ..., x_{MN}),$$

fixed columns

$${}^t(x_{00}, x_{10}, ..., x_{M0}), \quad {}^t(x_{0N}, x_{1N}, ..., x_{MN}).$$

The discrete variational principle is obtained using the variation of the first order of the action S.

Theorem 4.3: *The first variation of the discrete action S is*

$$\delta S(A)(\eta) = \frac{\partial L}{\partial x_{00}^i}(x_{00}, x_{10}, x_{01})\, \eta_{00}^i$$

$$+ \frac{\partial L}{\partial x_{10}^i}(x_{10}, x_{20}, x_{11})\, \eta_{10}^i + \frac{\partial L}{\partial x_{01}^i}(x_{01}, x_{11}, x_{02})\, \eta_{01}^i$$

$$+ \sum_{k=1}^{M-1} \sum_{\ell=1}^{N-1} \left[\frac{\partial L}{\partial x_{k\ell}^i}(x_{k\ell}, x_{k+1\ell}, x_{k\ell+1}) + \frac{\partial L}{\partial x_{k\ell}^i}(x_{k-1\ell}, x_{k\ell}, x_{k-1\ell+1}) \right.$$

$$\left. + \frac{\partial L}{\partial x_{k\ell}^i}(x_{k\ell-1}, x_{k+1\ell-1}, x_{k\ell}) \right] \eta_{k\ell}^i$$

$$+ \sum_{k=1}^{M-1} \frac{\partial L}{\partial x_{k0}^i}(x_{k-10}, x_{k0}, x_{k-11})\, \eta_{k0}^i + \sum_{k=0}^{M-1} \frac{\partial L}{\partial x_{kN}^i}(x_{kN-1}, x_{k+1N-1}, x_{kN})\, \eta_{kN}^i$$

$$+ \sum_{\ell=0}^{N-1} \frac{\partial L}{\partial x_{M\ell}^i}(x_{M-1\ell}, x_{M\ell}, x_{M-1\ell+1})\, \eta_{M\ell}^i + \sum_{\ell=1}^{N-1} \frac{\partial L}{\partial x_{0\ell}^i}(x_{0\ell-1}, x_{1\ell-1}, x_{0\ell})\, \eta_{0\ell}^i,$$

where

$$\eta = \begin{pmatrix} \eta_{00} & \eta_{01} & \cdots & \eta_{0N} \\ \eta_{10} & \eta_{11} & \cdots & \eta_{1N} \\ \cdots & \cdots & \cdots & \cdots \\ \eta_{M0} & \eta_{M2} & \cdots & \eta_{MN} \end{pmatrix}, \quad \eta_{k\ell} = \frac{\partial x_{k\ell}}{\partial \varepsilon}\Big|_{\varepsilon=0},$$

$$x_{k\ell}(\varepsilon) \in \mathbb{R}^n, \varepsilon \in I, 0 \in I, x_{k\ell}(0) = x_{k\ell}.$$

Proof. We use the family of curves $x_{k\ell}(\varepsilon) \in \mathbb{R}^n$, $\varepsilon \in I$. Then

$$S(A(\varepsilon)) = \sum_{k=0}^{M-1} \sum_{\ell=0}^{N-1} L(x_{k\ell}, x_{k+1\ell}, x_{k\ell+1}).$$

We obtain

$$\delta S(A)(\eta) = \frac{\partial}{\partial \varepsilon} S(A(\varepsilon))\Big|_{\varepsilon=0} = \sum_{k=1}^{M-1} \sum_{\ell=1}^{N-1} \frac{\partial L}{\partial x_{k\ell}^i}(x_{k\ell}, x_{k+1\ell}, x_{k\ell+1}) \eta_{k\ell}^i$$

$$+ \sum_{k=1}^{M-1} \sum_{\ell=1}^{N-1} \frac{\partial L}{\partial x_{k+1\ell}^i}(x_{k\ell}, x_{k+1\ell}, x_{k\ell+1}) \eta_{k+1\ell}^i$$

$$+ \sum_{k=1}^{M-1} \sum_{\ell=1}^{N-1} \frac{\partial L}{\partial x_{k\ell+1}^i}(x_{k\ell}, x_{k+1\ell}, x_{k\ell+1}) \eta_{k\ell+1}^i.$$

□

Corollary 4.1: **(Variational integrator)** *We denote*

$$D_1 L_d = \frac{\partial L_d}{\partial x_{k\ell}^i}(x_{k\ell}, x_{k+1\ell}, x_{k\ell+1}),$$

$$D_2 L_d = \frac{\partial L_d}{\partial x_{k\ell}^i}(x_{k-1\ell}, x_{k\ell}, x_{k-1\ell+1}),$$

$$D_3 L_d = \frac{\partial L_d}{\partial x_{k\ell}^i}(x_{k\ell-1}, x_{k+1\ell-1}, x_{k\ell}).$$

The matrix $A = (x_{k\ell})$ is stationary for the action S if and only if

$$D_1 L_d(x_{k\ell}, x_{k+1\ell}, x_{k\ell+1}) + D_2 L_d(x_{k-1\ell}, x_{k\ell}, x_{k-1\ell+1})$$

$$+ D_3 L_d(x_{k\ell-1}, x_{k+1\ell-1}, x_{k\ell}) = 0,$$

$i = \overline{1, n}; k = \overline{1, M-1}; \ell = \overline{1, N-1}.$

Proof. Since the boundary of the grid is fixed, the lines

$$(\eta_{00}, \eta_{01}, ..., \eta_{0N}), \ (\eta_{M0}, \eta_{M1}, ..., \eta_{MN})$$

and the columns

$${}^t(\eta_{00}, \eta_{10}, ..., \eta_{M0}), \ {}^t(\eta_{0N}, \eta_{M0}, \eta_{1N}, ..., \eta_{MN})$$

must be zero, while $\eta_{k\ell}, k = \overline{1, M-1}; \ell = \overline{1, N-1}$ are arbitrary. □
The transfer rule inside variational integrator is

$$(x_{k\ell-1}, x_{k+1\ell-1}, x_{k\ell}) \mapsto (x_{k-1\ell}, x_{k\ell}, x_{k-1\ell+1}) \mapsto (x_{k\ell}, x_{k+1\ell}, x_{k\ell+1}).$$

The variational integrator described by discrete Euler–Lagrange equations works as follows:

- we give the lines

$$(x_{00}, x_{01}, ..., x_{0N}), \quad (x_{10}, x_{11}, ..., x_{1N});$$

- we denote $u = x_{k\ell+1}$,

$$A_i(k\ell) = \frac{\partial L}{\partial x_{k\ell}^i}(x_{k-1,\ell}, x_{k\ell}, x_{k-1\ell+1})$$

$$B_i(k\ell) = \frac{\partial L}{\partial x_{k\ell}^i}(x_{k,\ell-1}, x_{k+1\ell-1}, x_{k\ell})$$

$$f_i(u) = \frac{\partial L}{\partial x_{k\ell}^i}(x_{k,\ell}, x_{k+1\ell}, u) + A_i(k\ell) + B_i(k\ell), \quad F = (f_1, ..., f_n);$$

- we solve the algebraic nonlinear system (2) $F(u) = 0$, at each step (k, ℓ) using six starting points as shown a part of the grid

$$
\begin{array}{ccc}
\clubsuit & \clubsuit x_{k-1\ell} & \clubsuit x_{k-1\ell+1} \\
\clubsuit x_{k\ell-1} & \clubsuit x_{k\ell} & * u = x_{k\ell+1} \\
\clubsuit x_{k+1\ell-1} & \clubsuit x_{k+1\ell} & \diamond
\end{array}
$$

The solution of the nonlinear system (2) can be approximated by using $u(\ell) = x_{k\ell+1}$ in the Newton method:

$$
J_F(u(e))
\begin{pmatrix}
u^1(e+1) \\
u^2(e+1) \\
... \\
u^n(e+1)
\end{pmatrix}
= J_F(u(e))
\begin{pmatrix}
u^1(e) \\
u^2(e) \\
... \\
u^n(e)
\end{pmatrix}
-
\begin{pmatrix}
f_1(u(e)) \\
f_2(u(e)) \\
... \\
f_n(u(e))
\end{pmatrix},
$$

for $e = 1, 2, ..., \bar{e}$. Here J_F means the Jacobi matrix of the function F.

A non-autonomous discrete Lagrangian L_d produces the discrete Hamiltonian

$$H_d(k, \ell) = \frac{x_{k+1\ell}^i - x_{k\ell}^i}{h_1} \frac{\partial L_d}{\partial x_1^i}$$

$$+ \frac{x_{k\ell+1}^i - x_{k\ell}^i}{h_2} \frac{\partial L_d}{\partial x_2^i} - L_d(kh_1, \ell h_2, x_{k\ell}, x_{k+1\ell}, x_{k\ell+1}).$$

Using H_d we can obtain the discrete version of Hamilton PDEs.

4.5 Numerical Study of the Vibrating String Motion

Instead of the continuous string, a discrete model is taken, consisting of a chain of n equal masses situated at equal distances along a flexible string, itself assumed to be massless.

The analysis of the vibrating string through the study of the relative discrete system is an effective teaching method, if one considers that the same conclusions can be reached in the analysis of continuous system, only by solving the differential equation of the waves, i.e., by using a mathematical tool which is not elementary.

The Lagrangian of a vibrating string is

$$L = \frac{1}{2} \left(\rho u_t^2(x,t) - \tau u_x^2(x,t) \right).$$

Using the step (h_1, h_2), we approximate

$$u_x \mapsto \frac{u_{k+1\ell} - u_{k\ell}}{h_1}, \quad u_t \mapsto \frac{u_{k\ell+1} - u_{k\ell}}{h_2}.$$

It follows the discretized Lagrangian

$$L_d(u_{k\ell}, u_{k+1\ell}, u_{k\ell+1}) = \frac{1}{2} \left(\rho \left(\frac{u_{k+1\ell} - u_{k\ell}}{h_1} \right)^2 - \tau \left(\frac{u_{k\ell+1} - u_{k\ell}}{h_2} \right)^2 \right).$$

Generally, a variational integrator with two indices is given by

$$D_1 L_d(u_{k\ell}, u_{k+1\ell}, u_{k\ell+1}) + D_2 L_d(u_{k-1\ell}, u_{k\ell}, u_{k-1\ell+1})$$
$$+ D_3 L_d(u_{k\ell-1}, u_{k+1\ell-1}, u_{k\ell}) = 0.$$

To build the variational integrator equations in case of vibrating string, we need the partial derivatives

$$D_1 L_d = \frac{-\rho}{h_1} \frac{u_{k+1\ell} - u_{k\ell}}{h_1} + \frac{\tau}{h_2} \frac{u_{k\ell+1} - u_{k\ell}}{h_2}$$

$$D_2 L_d = \frac{\rho}{h_1} \frac{u_{k\ell} - u_{k-1\ell}}{h_1}, \quad D_3 L_d = -\frac{\tau}{h_2} \frac{u_{k\ell} - u_{k\ell-1}}{h_2}.$$

Adding, we obtain the discrete equations

$$\frac{-\rho}{h_1} \frac{u_{k+1\ell} - u_{k\ell}}{h_1} + \frac{\tau}{h_2} \frac{u_{k\ell+1} - u_{k\ell}}{h_2}$$
$$+ \frac{\rho}{h_1} \frac{u_{k\ell} - u_{k-1\ell}}{h_1} + \frac{-\tau}{h_2} \frac{u_{k\ell} - u_{k\ell-1}}{h_2} = 0.$$

Simplifying, we get the variational integrator (bi-dimensional discrete system)

$$u_{k\ell+1} - 2u_{k\ell} + u_{k\ell-1} - \mathcal{C}^2(u_{k+1\ell} - 2u_{k\ell} + u_{k-1\ell}) = 0,$$

where

$$\mathcal{C} = \frac{h_2}{h_1} \sqrt{\frac{\rho}{\tau}}$$

is known as the *Courant number* (see paper [9]).

Remark 4.5: The *Courant number* is essential for the convergence condition by Courant–Friedrichs–Lewy theory. This is used for convergence while solving certain partial differential equations (usually hyperbolic PDEs) numerically. It arises in the numerical analysis of explicit time integration schemes, when these are used for the numerical solution. As a consequence, the time step must be less than a certain time in many explicit time-marching computer simulations, otherwise the simulation produces incorrect results.

Since the best solution is obtained for the value of $\mathcal{C}$ equal to one, we choose firstly the spatial step h_1 and then the temporal step h_2 such that $\mathcal{C} = 1$. We will solve this discrete two-dimensional system in three situations (three problems on the vibrating string): initial conditions, initial and boundary conditions, and soliton solutions.

4.5.1 Initial Conditions for Infinite String

For infinite string, we have $t \geq 0$. The initial conditions are $u(x,0) = f(x)$, $\frac{\partial u}{\partial t}(x,0) = g(x)$, whence

$$u_{k0} = f(x_k), \quad \frac{u_{k1} - u_{k0}}{h_2} = g(x_k), \quad \text{for } k \in \mathbb{Z}.$$

We find $u_{k1} = h_2\, g(x_k) + u_{k0}$. Now we can compute

$$u_{k2} = \mathcal{C}^2(u_{k+11} - 2u_{k1} + u_{k-11}) + 2u_{k1} - u_{k0}, \quad \text{for } k \in \mathbb{Z},$$

by variational integrator, and so on.

More precisely, for $\mathcal{C} = 1$, the variational integrator becomes

$$u_{k\,\ell+1} + u_{k\,\ell-1} = u_{k+1\,\ell} + u_{k-1\,\ell}, \quad k \in \mathbb{Z},\ \ell \in \mathbb{N}.$$

We obtain

$$u_{k\,\ell+1} - u_{k+1\,\ell} = u_{k-1\,\ell} - u_{k\,\ell-1} = \cdots = u_{k-\ell\,1} - u_{k-\ell+1\,0}$$

and

$$u_{k\,\ell+1} - u_{k-1\,\ell} = u_{k+1\,\ell} - u_{k\,\ell-1} = \cdots = u_{k+\ell\,1} - u_{k+\ell+1\,0}.$$

Remark 4.6: It may be found that $u_{k\ell}$ depends upon $k - \ell$ and $k + \ell$ only.

Further we compute

$$u_{k\ell} = u_{k-\ell\,0} + \sum_{j=0}^{j=\ell-1} (u_{k-\ell+j+1\,j+1} - u_{k-\ell+j\,j})$$

$$= u_{k-\ell\,0} + \sum_{j=0}^{j=\ell-1} (u_{k-\ell+2j+1\,1} - u_{k-\ell+2j\,0}).$$

If h_1 goes to zero, the prescribed values u_{k0} converge uniformly to the twice continuously differentiable function, $f(x)$, and the difference quotients $\dfrac{u_{k1} - u_{k0}}{h_2\sqrt{2}}$ there converge uniformly to a continuously differentiable function $g(x)$, then the last side of the above relation goes uniformly to

$$u(x,t) = f(x-t) + \frac{1}{\sqrt{2}} \int_{x-t}^{x+t} g(s)\,ds.$$

This is the well-known expression for the solution of the vibrating string equation $\frac{\partial^2 u}{\partial t^2} = \frac{\partial^2 u}{\partial x^2}$ with initial conditions $u(x,0) = f(x)$ and $\frac{\partial u}{\partial t}(x,0) = -f'(x) + \sqrt{2}g(x)$. This shows that as $h_1 \to 0$ and $h_2 \to 0$, the solution of the variational integrator converges to the solution of the PDE provided the initial values converge appropriately (as above).

The situation $\mathcal{C} \neq 1$ is more complicated and we send the reader to the paper [9], the first to address such issues.

4.5.2 Finite String, Fixed at the Ends

Let us solve the variational integrator system in case of the finite string, length L, fixed at the ends. Once more $\mathcal{C} = 1$. The boundary conditions are $u(0,t) = u(L,t) = 0 \ \forall t \in \mathbb{R}$, whence $u_{0\ell} = u_{N\ell} = 0$ for $\ell \in \mathbb{N}$. To find a formula for the general term $u_{k\ell}$ we make the following considerations.

Imitating the method of separating the variables let us look for solutions of the form $u_{k\ell} = u_k v_\ell$. Then we can split the previous equation as

$$\frac{u_{k+1} + u_{k-1}}{u_k} = \frac{v_{\ell+1} + v_{\ell-1}}{v_\ell} = 2\lambda,$$

or as the pair of discrete equations

$$u_{k+1} - 2\lambda u_k + u_{k-1} = 0,$$
$$v_{\ell+1} - 2\lambda v_\ell + v_{\ell-1} = 0,$$

where $\lambda \in \mathbb{R}$ is to be determined from boundary conditions.

Remark 4.7: The above type of relation is the recurrence relation for Chebyshev's polynomials in variable λ.

In the case of the finite string, fixed at the ends, the boundary conditions are $u(0,t) = u(L,t) = 0 \ \forall t$, whence $u_0 = u_N = 0$. The solution for the sequence $\{u_k\}$ has the form $u_k = C_1 r_1^k + C_2 r_2^k$, where r_1, r_2 are the roots of the characteristic equation $r^2 - 2\lambda r + 1 = 0$. In order for the conditions to be verified at the ends, the characteristic equation must have imaginary roots, that is $\lambda^2 - 1 < 0$, or $\lambda \in (-1,1)$, and then it follows $r_{1,2} = \lambda \pm i\sqrt{1-\lambda^2} = \cos\theta \pm i\sin\theta$, where $\cos\theta = \lambda$, because $|r_{1,2}| = 1$. Hence $u_k = C_1 \cos k\theta + C_2 \sin k\theta$, $k = \overline{0,N}$ and the boundary conditions impose $C_1 = 0$, and $\theta = \frac{n\pi}{N}$, $n = \overline{1,N-1}$, i.e. $\lambda_n = \cos\frac{n\pi}{N}$. Then we have

$$(u_k)_n = C_n \sin\frac{n\pi}{N}k, \quad k = \overline{0,N}, \ n = \overline{1,N-1}.$$

Analogously, with λ previously determined, the equation for v_ℓ becomes

$$v_{\ell+1} - 2\cos\frac{n\pi}{N}v_\ell + v_{\ell-1} = 0.$$

One obtains

$$(v_\ell)_n = A_n \cos\frac{\pi n}{N}\ell + B_n \sin\frac{\pi n}{N}\ell,$$

and finally

$$u_{k\ell} = \sum_{n=1}^{N-1}\left(A_n \cos\frac{\pi n\ell}{N} + B_n \sin\frac{\pi n\ell}{N}\right)\sin\frac{\pi nk}{N}.$$

The $2(N-1)$ coefficients A_n and B_n will be determined from the $2(N-1)$ values for u_{k0} and u_{k1} deduced, as above, from the initial conditions.

Remark 4.8: Taking into account that $L = Nh_1$, we emphasize the spatial component having $(u_k)_n = C_n \sin\frac{n\pi}{L}kh_1 = C_n \sin\frac{n\pi}{L}x_k$. If $h_1 \rightarrow 0$, the value of $u_{k\ell}$ tends to

$$u(x,t) = \sum_{n=1}^{\infty}\left(A_n \cos\frac{\pi nt}{L} + B_n \sin\frac{\pi nt}{L}\right)\sin\frac{\pi nx}{L},$$

well-known expression of the solution of the finite vibrating string equation, fixed at the ends.

The problem for the wave equation with initial conditions and boundary conditions has unique solution.

4.5.3 Monomial (Soliton) Solutions

We start with the bi-dimensional discrete system (variational integrator)

$$\frac{\rho}{h_1^2}(2u_{k\ell} - u_{k-1\ell} - u_{k+1\ell}) + \frac{\tau}{h_2^2}(u_{k\ell+1} - 2u_{k\ell} + u_{k\ell-1}) = 0.$$

Denoting

$$a_1 = \frac{\rho}{h_1^2},\ a_2 = \frac{\tau}{h_2^2},$$

we find the bi-dimensional discretized system

$$2(a_1 - a_2)u_{k\ell} - a_1(u_{k-1\ell} + u_{k+1\ell}) + a_2(u_{k\ell+1} + u_{k\ell-1}) = 0.$$

Let us determine the conditions in which the equation (k, ℓ) has the monomial solution

$$u_{k\ell} = \alpha\, r_1^k r_2^\ell,\ \alpha \neq 0.$$

It follows

$$2(a_1 - a_2)\alpha r_1^k r_2^\ell - a_1(\alpha r_1^{k-1}r_2^\ell + \alpha r_1^{k+1}r_2^\ell) + a_2(\alpha r_1^k r_2^{\ell+1} + \alpha r_1^k r_2^{\ell-1}) = 0$$

and, simplifying by $\alpha r_1^{k-1} r_2^{\ell-1}$, we find the characteristic equation (characteristic curve in $\mathbb{R}^2$, see Fig. 4.1)

$$2(a_1 - a_2)r_1 r_2 - a_1 r_2(1 + r_1^2) + a_2 r_1(1 + r_2^2) = 0$$

or

$$a_1 r_2(r_1 - 1)^2 - a_2 r_1(r_2 - 1)^2 = 0.$$

The singular point $(1, 1)$ belongs to the characteristic curve, and hence this curve is not a void set.

The characteristic equation is determined by the coefficients a_1, a_2. This means it is the same for all equations indexed by (k, ℓ). The characteristic equation is invariant with respect to the diagonal translation $u_{k\ell} \mapsto u_{k+1\ell+1}$. Also, the characteristic curve is a rational one with parameterization

$$r_1 = \frac{a_1(\mu - 1)}{a_1\mu - a_2\mu^2}, \quad r_2 = \frac{a_2\mu^2(\mu - 1)}{a_1\mu - a_2\mu^2}, \quad \mu \in \mathbb{R} \setminus \{0, a_1/a_2\}.$$

This curve is not included in the gallery of cubic plane curves. Because this curve appears for the first time in this type of study, we call it the Udriste curve.

We add a Maple program (Fig. 4.1):

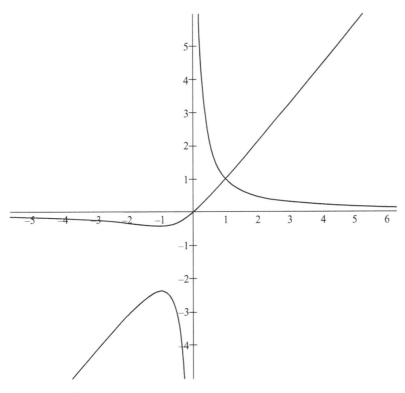

Figure 4.1 Characteristic curve.

> $rho := 12; tau := 10; h1 := 2.; h2 := 2.; r2 := -1; \alpha := .5$
> $a1 := \rho/h1^2; a2 := \tau/h2^2$
> with(plots, implicitplot):
> $implicitplot(a1 * y * (x - 1)^2 - a2 * x * (y - 1)^2 = 0, x = -2..2, y = -2..2, grid = [25, 25], scaling = constrained);$

Remark 4.9: For $\mathcal{C} = 1$, the characteristic equation becomes $(r_1 - r_2)(r_1 r_2 - 1) = 0$ and the characteristic curve decomposes into the first bisector of the axes $r_1 - r_2 = 0$ and the equilateral hyperbola $r_1 r_2 = 1$.

We set $r_2 > 0$ as a parameter and order for unknown r_1 (see explicit Cartesian representation of characteristic curve):

$$a_1 r_2 r_1^2 - (a_2 r_2^2 - 2(a_1 - a_2)r_2 + a_2)r_1 + a_1 r_2 = 0.$$

It follows two real solutions r_{11} and r_{12}.
    We add another Maple program.
> $rho := 9; tau := 2; h1 := .3; h2 := .2; r2 := .8; \alpha := 5;$
> $a1 := rho/h1^2; a2 := \tau/h2^2;$
> $eq1 := a1 * r2 * x^2 - (a2 * (r2 - 1)^2 + 2 * a1 * r2) * x + a1 * r2$
> $solutions := [solve(eq1, x)]$
> $k := 10; l := 25; u(k, l) = \alpha * r11^k * r2^l$
> $m := 40; n := 35$
The graph of the function $z = \alpha * r11^{x1} * r2^{x2}$ for $r1 = r11$ (see Fig. 4.2)
> $plot3d(\alpha * r11^{x1} * r2^{x2}, x1 = 0..m, x2 = 0..n)$

Figure 4.2 Graph of the function z.

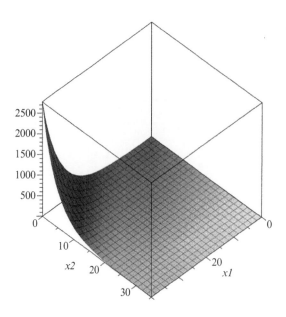

Theorem 4.4:   *The monomial expressions* $u_{k\ell} = \alpha\, r_1^k r_2^\ell$, $\alpha \in \mathbb{R}$ *may be solutions of the equation* (k,ℓ) *with initial condition* $u_{00} = c_1$ *and final condition* $u_{NM} = c_2$, *for* $c_1 c_2 \neq 0$.

Proof.   It follows $\alpha = c_1$. The final condition gives $r_1^N r_2^M = \dfrac{c_2}{c_1}$, relation which together with the characteristic equation give us r_1 and r_2.

For instance, if $\mathcal{C} = 1$, then one obtains $r_1 = r_2 = \left(\dfrac{c_2}{c_1}\right)^{\frac{1}{M+N}}$, or $r_1 = \left(\dfrac{c_2}{c_1}\right)^{\frac{1}{N-M}}$ and $r_2 = \left(\dfrac{c_2}{c_1}\right)^{\frac{1}{M-N}}$.

Theorem 4.5:   *Suppose* r_1, r_2 *are constrained by the characteristic equation. If each* (k,ℓ) *discrete equation has the solution* $u_{k\ell} = \alpha_{k\ell}\, r_1^k r_2^\ell$ *(no sum), then* $u_{k'\ell'} = \sum_{k,\ell=1}^{n} \alpha_{k'\ell'k\ell}\, r_1^k r_2^\ell$ *(double sum, sygnomial solution) is the general solution (finite series approximation) of the vibrating string integrator system.*

Proof.   Each term of the sum is solution of each (k,ℓ) discrete equation.   □

Remark 4.10:   For the continuous wave equation, monomial solutions become soliton-type solutions $u(x,t) = \alpha e^{ax+\omega t}$.

Remark 4.11:   A similar study can be done for the solution of the discretization of a linear PDE with two independent variables and constant coefficients.

Remark 4.12:   The solution of the birecurrent equation can be of the type

$$u(k,\ell) = \alpha f(r_1)^k f(r_2)^\ell,$$

for a given function f. One of the values r_1 or r_2 is given and the other is determined from the Udriste equation

$$a_1 \frac{(f(r_1) - 1)^2}{f(r_1)} = a_2 \frac{(f(r_2) - 1)^2}{f(r_2)}.$$

Case simulation, $f(x) = \sin(x)$, via Maple.

```
> r2 := (2 * π)/3;
> g := (sin(r2) − 1)^2/sin(r2);
> m := (2 * a1 + a2 * g − sqrt((a2 * g + 2 * a1)^2 − 4 * a1^2))/(2 * a1);
> with(plots, implicitplot):
> p := sin(r2);
> α := 0.5;
> m := 2;
> n := 3;
> plot3d(α * p^x1 * m^x2, x1 = −1..m, x2 = −2..n);
```

The most general case that works is: $u(k, \ell) = \alpha f_1(r_1)^k f_2(r_2)^\ell$. The pairs $f_1(x) = \sin(x), f_2(x) = \cos(x)$ and $f_1(x) = x, f_2(x) = \sin(x)$ can be detailed immediately. Theorems similar to Theorems 4.4, 4.5 can be formulated in this cases as well.

Remark 4.13: The discretization of the wave equation

$$u_{tt}(x, t) = a^2 u_{xx}(x, t),$$

obtained through the correspondences

$$u_{xx} \mapsto \frac{u_{k+1\ell} - 2u_{k\ell} + u_{k-1\ell}}{h_1^2}, \quad u_{tt} \mapsto \frac{u_{k\ell+1} - 2u_{k\ell} + u_{k\ell-1}}{h_2^2}$$

coincides with the previous integrator, modulo the coefficients that must be arranged for identification.

4.5.4 More About Recurrence Relations

Let us consider the bi-dimensional discrete system (variational integrator)

$$a_1(2u_{k\ell} - u_{k-1\ell} - u_{k+1\ell}) + a_2(u_{k\ell+1} - 2u_{k\ell} + u_{k\ell-1}) = 0,$$

where

$$a_1 = \frac{\rho}{h_1^2}, \quad a_2 = \frac{\tau}{h_2^2}.$$

Let us look for solutions of the decomposed form $u_{k\ell} = u_k v_\ell$. Then we can write the previous discrete equation as

$$a_1 \frac{u_{k+1} - 2u_k + u_{k-1}}{u_k} = a_2 \frac{v_{\ell+1} - 2v_\ell + v_{\ell-1}}{v_\ell} = \lambda,$$

or as the pair of equations

$$a_1 u_{k+1} - (2a_1 + \lambda)u_k + a_1 u_{k-1} = 0$$

$$a_2 v_{\ell+1} - (2a_2 + \lambda)v_\ell + a_2 v_{\ell-1} = 0,$$

where the parameter $\lambda \in \mathbb{R}$ is to be determined from boundary conditions.

Let us study the second-order linear difference equation

$$a_1 u_{k+1} - (2a_1 + \lambda)u_k + a_1 u_{k-1} = 0.$$

The only stationary solution to this difference equation is $u_k = 0$. For the general solution, let us look at the roots of a new characteristic (reciprocal) equation $a_1 r^2 - (2a_1 + \lambda)r + a_1 = 0$, i.e.,

$$r_{1,2} = \frac{2a_1 + \lambda \pm \sqrt{4a_1\lambda + \lambda^2}}{2a_1}, \quad \Delta = 4a_1\lambda + \lambda^2.$$

If $\Delta > 0$ we get two real roots while if $\Delta < 0$ we have a pair of complex roots.

Although in this particular case $r_1 r_2 = 1$, we will further discuss the general case to which we will apply this particularity:

(i) $\Delta > 0$. Then, a solution $\{u_k\}$ has the form $u_k = C_1(a_1, \lambda) r_1^k + C_2(a_1, \lambda) r_2^k$, where C_1, C_2 are two as yet undetermined constants. If $|r_1|$ or $|r_2| > 1$, the values of u_k will diverge to $\pm\infty$, oscillating or not. Otherwise, u_k will converge to 0 exponentially.

In our particular case $r_2 = 1/r_1 \neq 1$ and the solution is $u_k = C_1(a_1, \lambda) r_1^k + C_2(a_1, \lambda) r_1^{-k}$. Because $|r_1|$ or $|1/r_1| > 1$, the values of u_k will diverge to $\pm\infty$, oscillating or not.

(ii) $\Delta < 0$. The roots will be complex conjugated: $r_1 = \rho e^{i\theta}$ and $r_2 = \rho e^{-i\theta}$. Then $u_k = C_1(a_1, \lambda)\rho^k e^{i\theta k} + C_2(a_1, \lambda)\rho^k e^{-i\theta k}$. Since u_k is real for all k, the constants C_1 and C_2 must be complex conjugated. For example, $u_k = C_1\rho e^{i\theta} + C_2\rho e^{-i\theta} = \rho(C_1 + C_2)\cos\theta + i\rho(C_1 - C_2)\sin\theta$ is real. Therefore, $C_1 + C_2$ must be real and $C_1 - C_2$ purely imaginary, so that if $C_1 = x + iy$ we must have $C_2 = x - iy$. Let $C_1 = ae^{i\phi}$ and $C_2 = ae^{-i\phi}$. Then we can write $u_k = 2a\rho^k \cos(\theta k + \phi)$. Here u_k will be a spiral, in the plane (ρ, θ). In fact we can get almost any continuous function of x by taking sums of sine and cosine curves (this is the idea behind Fourier representation theory). If $|\rho| > 1$, then the amplitude of the oscillations will explode exponentially. If $|\rho| < 1$, the amplitude will decay to zero.

In our case $|\rho| = 1$, and we are to get a cycle of the same amplitude continuing forever. This is the situation in Subsection 4.5.2.

(iii) $\Delta = 0$. Here $r_1 = r_2 = r$ and the solution to the difference equation is of the form $u_k = r^k (C_1 + C_2 k)$. Then u_k will follow a linear trend, multiplied by an exploding or decaying exponential depending on whether $|r| > 1$ or $|r| < 1$.

In our case $r = \pm 1$ and u_k will follow a linear trend, oscillating or not.

For the recurrence in v_ℓ, we have similar results.

4.5.5 Solution by Maple via Eigenvalues

To solve the recurrence $a_1 u_{k+1} - (2a_1 + \lambda)u_k + a_1 u_{k-1} = 0$, we use the Maple command

> rsolve(a1 $*$ f(n + 1) − (2 $*$ a1 + λ) $*$ f(n)
+a1 $*$ f(n − 1) = 0, f(0) = 0, f(10) = 0, f(k))

To simplify, we denote

$$A = (11a_1^{10} + 220a_1^9\lambda + 1287a_1^8\lambda^2 + 3432a_1^7\lambda^3 + 5005a_1^6\lambda^4$$
$$+ 4368a_1^5\lambda^5 + 2380a_1^4\lambda^6 + 816a_1^3\lambda^7 + 171a_1^2\lambda^8 + 20a_1\lambda^9$$
$$+ \lambda^{10})\sqrt{\Delta};$$

$$B = 2a_1^{11} + 121a_1^{10}\lambda + 1210a_1^9\lambda^2 + 4719a_1^8\lambda^3 + 9438a_1^7\lambda^4$$
$$+ 11011a_1^6\lambda^5 + 8008a_1^5\lambda^6 + 3740a_1^4\lambda^7 + 1122a_1^3\lambda^8 + 209a_1^2\lambda^9$$
$$+ 22a_1\lambda^{10} + \lambda^{11}.$$

It follows $f(k) = u_k = C_1(a_1, \lambda) r_1^{k+1} + C_2(a_1, \lambda) r_2^{k+1}$, where

$$C_1(a_1, \lambda) = \frac{f(9)}{2a_1^{10}\sqrt{\Delta}} (A - B), \quad C_2(a_1, \lambda) = \frac{f(9)}{2a_1^{10}\sqrt{\Delta}} (A + B).$$

This solution depends on the value $f(9)$.

For the recurrence in v_ℓ, we find a similar solution. Then $u_{k\ell} = u_k v_\ell$ is a solution, and for the general solution we can apply the Theorem 4.5.

4.5.6 Solution by Maple via Matrix Techniques

We denote $x(i) = f(i) . i = 0..7. v1 = [-a1 * u0, 0, 0, 0, 0, 0, -a1 * u8]$; $A1$ band matrix, $A1 * x = v1$ the system which determines $f(i)$, the matrix $B1$ is the inverse of $A1$; graphs of the functions $f(i)$ in relation to λ are made. The solution is unique.

We use another Maple command

```
> restart;
> with(linalg);
> a1 := 2;
> b1 := 2 * a1 + λ;
> A1 := matrix(7, 7, [-b1, a1, 0, 0, 0, 0, 0, a1, -b1, a1,
0, 0, 0, 0, 0, a1, -b1, a1, 0, 0, 0, 0, 0, a1, -b1, a1, 0, 0,
0, 0, 0, a1, -b1, a1, 0, 0, 0, 0, 0, a1, -b1, a1, 0, 0, 0, 0, 0, a1, -b1]);
> A1 := [[-4 - λ, 2, 0, 0, 0, 0, 0],
[2, -4 - λ, 2, 0, 0, 0, 0],
[0, 2, -4 - λ, 2, 0, 0, 0],
[0, 0, 2, -4 - λ, 2, 0, 0],
[0, 0, 0, 2, -4 - λ, 2, 0],
[0, 0, 0, 0, 2, -4 - λ, 2],
[0, 0, 0, 0, 0, 2, -4 - λ]]
> B1 := inverse(A1)
```

Case 1: $u_0 = 0$; $u_8 = 2$;

```
> v1 := vector(7, [-a1 * u0, 0, 0, 0, 0, 0, -a1 * u8])
> x := multiply(B1, v1)
> x1 := x[1]; x2 := x[2]; x3 := x[3]; x4 := x[4]; x5 := x[5]; x6 := x[6]; x7 := x[7]
```

$> x1 := 256/(\lambda^7 + 28 * \lambda^6 + 312 * \lambda^5 + 1760 * \lambda^4 + 5280 * \lambda^3 + 8064 * \lambda^2 + 5376 * \lambda + 1024)$

$> x2 := 128/(\lambda^6 + 24 * \lambda^5 + 216 * \lambda^4 + 896 * \lambda^3 + 1696 * \lambda^2 + 1280 * \lambda + 256)$

$> x3 := (64 * (\lambda^2 + 8 * \lambda + 12))/(\lambda^7 + 28 * \lambda^6 + 312 * \lambda^5 + 1760 * \lambda^4 + 5280 * \lambda^3 + 8064 * \lambda^2 + 5376 * \lambda + 1024)$

$> x4 := 32/(\lambda^4 + 16 * \lambda^3 + 80 * \lambda^2 + 128 * \lambda + 32)$

$> x5 := (16 * (\lambda^4 + 16 * \lambda^3 + 84 * \lambda^2 + 160 * \lambda + 80))/(\lambda^7 + 28 * \lambda^6 + 312 * \lambda^5 + 1760 * \lambda^4 + 5280 * \lambda^3 + 8064 * \lambda^2 + 5376 * \lambda + 1024)$

$> x6 := (8 * (\lambda^4 + 16 * \lambda^3 + 80 * \lambda^2 + 128 * \lambda + 48))/(\lambda^6 + 24 * \lambda^5 + 216 * \lambda^4 + 896 * \lambda^3 + 1696 * \lambda^2 + 1280 * \lambda + 256)$

$> x7 := (4 * (\lambda^6 + 24 * \lambda^5 + 220 * \lambda^4 + 960 * \lambda^3 + 2016 * \lambda^2 + 1792 * \lambda + 448))/(\lambda^7 + 28 * \lambda^6 + 312 * \lambda^5 + 1760 * \lambda^4 + 5280 * \lambda^3 + 8064 * \lambda^2 + 5376 * \lambda + 1024)$

```
> a1 := 1;
> plot(x1, λ = -6..6); see Fig. 4.3
> plot(x2, λ = -6..6); see Fig. 4.4
```

We find a similar solution for the recurrence in v_ℓ. Then $u_{k\ell} = u_k v_\ell$ is a solution, and for the general solution we can apply the Theorem 4.5.

Figure 4.3 Graph of the function x1.

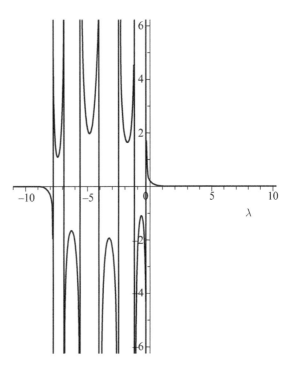

Figure 4.4 Graph of the function x2.

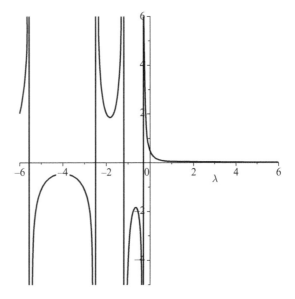

4.6 Numerical Study of the Vibrating Membrane Motion

In this section we are interested in obtaining an approximated numerical solution for the model of vibrating elastic membranes.

Let $u = u(x, y; t)$ be the function that represents the small transverse displacement of the point (x, y) in the membrane. The Lagrangian of a vibrating membrane is

$$L = \frac{1}{2} \left(\rho u_t^2 - \tau \left(u_x^2 + u_y^2 \right) \right).$$

Let us write the discrete version of the Lagrangian L and realize a numerical study of the motion of a vibrating membrane.

We use the 3-step (h_1, h_2, h_3) and the approximations

$$u_x \mapsto \frac{u_{h+1k\ell} - u_{hk\ell}}{h_1}, \quad u_y \mapsto \frac{u_{hk+1\ell} - u_{hk\ell}}{h_2}, \quad u_t \mapsto \frac{u_{hk\ell+1} - u_{hk\ell}}{h_3}.$$

It follows the discretized Lagrangian

$$2L_d(u_{hk\ell}, u_{h+1k\ell}, u_{hk+1\ell}, u_{hk\ell+1})$$

$$= \rho \left(\frac{u_{hk\ell+1} - u_{hk\ell}}{h_3} \right)^2 - \tau \left(\left(\frac{u_{h+1k\ell} - u_{hk\ell}}{h_1} \right)^2 + \left(\frac{u_{hk+1\ell} - u_{hk\ell}}{h_2} \right)^2 \right).$$

Generally, the variational integrator has the equations

$$D_1 L_d(u_{hk\ell}, u_{h+1k\ell}, u_{hk+1\ell}, u_{hk\ell+1})$$

$$+ D_2 L_d(u_{h-1k\ell}, u_{hk\ell}, u_{h-1k+1\ell}, u_{h-1k\ell+1})$$

$$+ D_3 L_d(u_{hk-1\ell}, u_{h+1k-1\ell}, u_{hk\ell}, u_{hk-1\ell})$$

$$+ D_4 L_d(u_{hk\ell-1}, u_{h-1k\ell-1}, u_{hk+1\ell-1}, u_{hk\ell}) = 0.$$

In our case, we find

$$D_1 L_d = \left(\frac{\rho}{h_3^2} - \frac{\tau}{h_1^2} - \frac{\tau}{h_2^2} \right) u_{hk\ell} - \frac{\rho}{h_3^2} u_{hk\ell+1} + \frac{\tau}{h_1^2} u_{h+1k\ell} + \frac{\tau}{h_2^2} u_{hk+1\ell},$$

$$D_2 L_d = -\frac{\tau}{h_1^2} (u_{hk\ell} - u_{h-1k\ell}), \quad D_3 L_d = -\frac{\tau}{h_2^2} (u_{hk\ell} - u_{hk-1\ell}),$$

$$D_4 L_d = \frac{\rho}{h_3^2} (u_{hk\ell} - u_{hk\ell-1}).$$

Denote $a_1 = \frac{\tau}{h_1^2}$, $a_2 = \frac{\tau}{h_2^2}$, $a_3 = \frac{\rho}{h_3^2}$. The variational integrator (three-dimensional discretized system) is

$$2(a_3 - a_1 - a_2) u_{hk\ell} + a_1(u_{h+1k\ell} + u_{h-1k\ell})$$

$$+ a_2(u_{hk+1\ell} + u_{hk-1\ell}) - a_3(u_{hk\ell+1} + u_{hk\ell-1}) = 0.$$

4.6.1 Monomial (Soliton) Solutions

Let us look for solutions of monomial type,

$$u_{hk\ell} = \alpha \, r_1^h r_2^k r_3^\ell, \ \alpha > 0,$$

for each equation (h, k, ℓ). It follows the characteristic equation (characteristic surface in $\mathbb{R}^3$, Figs 4.5 and 4.6)

$$2(a_3 - a_1 - a_2)r_1 r_2 r_3 + a_1 r_2 r_3(r_1^2 + 1) + a_2 r_1 r_3(r_2^2 + 1) - a_3 r_1 r_2(r_3^2 + 1) = 0$$

or

$$a_1 r_2 r_3(r_1 - 1)^2 + a_2 r_3 r_1(r_2 - 1)^2 - a_3 r_1 r_2(r_3 - 1)^2 = 0.$$

The singular point $(1, 1, 1)$ belongs to the characteristic surface, and hence this surface is not a void set.

The characteristic equation is determined only by the coefficients a_1, a_2, a_3. This means, it is the same for all equations indexed by (h, k, ℓ). The characteristic equation is invariant with respect to the diagonal translation $u_{hk\ell} \mapsto u_{h+1k+1\ell+1}$.

Because this surface is not included in the gallery of algebraic surfaces, and Constantin Udriste was the first detailing this type of study, we call it the Udriste surface.

We add a Maple program.

> with(plots);

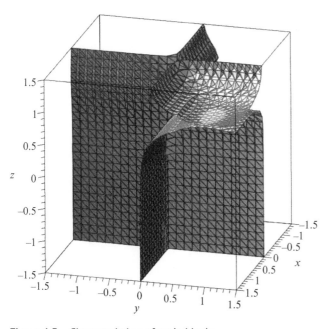

Figure 4.5 Characteristic surface in Maple.

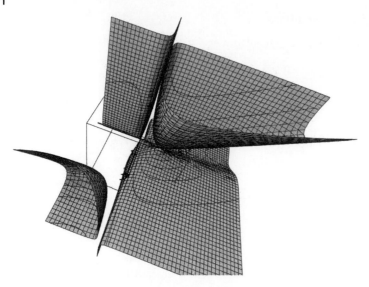

Figure 4.6 Characteristic surface in CalcPlot3D.

> $implicit\,plot3d(a1 * y * z * (x-1)^2 + a2 * z * x * (y-1)^2 - a3 * x * y * (z-1)^2 = 0, x = -1.5..1.5, y = -1.5..1.5, z = -1.5..1.5, grid = [25, 25, 25])$;

Keeping fixed two of the unknowns r_1, r_2, r_3, we find real solutions in relation to the third chosen unknown (explicit Cartesian representation of characteristic surface).

For Fig. 4.7, we need another Maple program.

> $rho := 2; tau := 2; h1 := 2.; h2 := 1.; h3 := .2; r1 := 2; r2 := 1.; alpha := .5;$
> $a1 := rho/h1^2; a2 := tau/h2^2; a3 := tau/h3^2;$
> $eq1 := a3 * r1 * r2.(x^2) - ((2 * (a3 - a1 - a2)) * r1 * r2 + a1 * r2 * (r1^2 + 1) + a2 * r1 * (r2^2 + 1)) * x + a3 * r1 * r2;$
> $solutions := [solve(eq1, x)];$
> $r31 := solutions[1]; r32 := solutions[2];$
> $k := 10; h := 2; l := 25; u(k, l) = alpha * r1^h * r2^k * r31^l;$
> $m := 5; n := 10; x2 := 5;$

The graph of the function $u = alpha * r1^{x1} * r2^{x2} * r31^{x3}$ for fixed $x2$ (see Fig. 4.7)

> $plot3d(alpha * r1^{x1} * r2^{x2} * r31^{x3}, x1 = 0..m, x3 = 0..n);$

Theorem 4.6: *Suppose the numbers r_1, r_2, r_3 are constrained by the characteristic equation. If each (h, k, ℓ) discrete equation has the solution $u_{hk\ell} = \alpha_{hk\ell}\, r_1^h r_2^k r_3^\ell$ (no sum), then*

$$u_{h'k'\ell'} = \sum_{h,k,\ell=1}^{n} \alpha_{h'k'\ell'hk\ell}\, r_1^h r_2^k r_3^\ell \quad \text{(triple sum, sygnomial)}$$

is the general solution (finite series approximation) of the vibrating membrane integrator system.

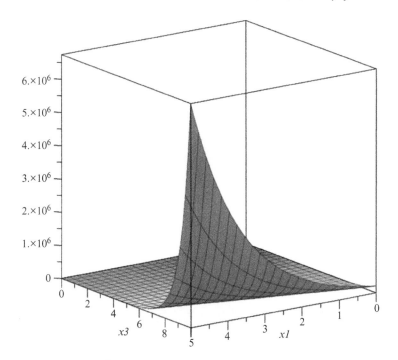

Figure 4.7 Graph of the function u for fixed $x2$.

Proof. Each term of the sum is the solution of each (h, k, ℓ) discrete equation. $\qquad\square$

Remark 4.14: A similar study can be done for the solution of the discretization of certain linear PDE with three independent variables and constant coefficients.

Remark 4.15: The discretization of the bi-dimensional wave equation

$$u_{tt}(x, y, t) = \alpha^2(u_{xx}(x, y, t) + u_{yy}(x, y, t)),$$

obtained through the correspondences

$$u_{xx} \mapsto \frac{u_{h+1k\ell} - 2u_{hk\ell} + u_{h-1k\ell}}{h_1^2},$$

$$u_{yy} \mapsto \frac{u_{hk+1\ell} - 2u_{hk\ell} + u_{hk-1\ell}}{h_2^2},$$

$$u_{tt} \mapsto \frac{u_{hk\ell+1} - 2u_{hk\ell} + u_{hk\ell-1}}{h_3^2},$$

coincides with the previous variational integrator, modulo the coefficients that must be arranged for identification.

4.6.2 Initial and Boundary Conditions

Let us solve the variational integrator system in case of the rectangular membrane fixed on the boundary. We fix the boundary conditions $u(0, y, t) = u(a, y, t) = 0; u(x, 0, t) = u(x, b, t) = 0 \ \forall t \in \mathbb{R}$, whence $u_{0k\ell} = u_{Mk\ell} = u_{h0\ell} = u_{hN\ell} = 0$. The initial conditions are $u(x, y, 0) = f(x, y)$, $\frac{\partial u}{\partial t}(x, y, 0) = g(x, y)$, whence

$$u_{hk0} = f(x_h, y_k), \quad \frac{u_{hk1} - u_{hk0}}{h_3} = g(x_h, y_k),$$

for $h = \overline{1, M-1}, \ k = \overline{1, N-1}$. From here we find $u_{hk1} = h_3\, g(x_h, y_k) + u_{hk0}$. Now we can compute by variational integrator

$$u_{hk2} = -\frac{1}{a_3}\left[2(a_3 - a_1 - a_2)u_{hk1} + a_1(u_{h+1k1} + u_{h-1k1})\right.$$

$$\left. + a_2(u_{hk+11} + u_{hk-11}) - u_{hk0}\right], \quad \text{for} \ \ h = \overline{1, M-1}, \ k = \overline{1, N-1},$$

and so on.

Remark 4.16: The same procedure is applicable if the restrictions on the boundary of the rectangular domain are nonzero, i.e. $u(x, 0, t) = \varphi_1(x, t)$, $u(x, b, t) = \varphi_2(x, t)$; $u(0, y, t) = \psi_1(y, t)$, $u(a, y, t) = \psi_2(y, t)$, or for an infinite membrane.

As in case of the vibrating string firstly, for simplicity, we take $h_1 = h_2$ and h_3 such that $a_3 = a_1 + a_2$ and secondly we can write the variational integrator in the form

$$(u_{h-1k\ell} - 2u_{hk\ell} + u_{h+1k\ell}) + (u_{hk-1\ell} - 2u_{hk\ell} + u_{hk+1\ell})$$

$$= 2(u_{hk\ell-1} - 2u_{hk\ell} + u_{hk\ell+1}).$$

This form suggests to use the method of separating the variables. Let us look for solutions of the form $u_{hk\ell} = u_h v_k w_\ell$. Then we can write the previous equation as

$$\frac{u_{h-1} + u_{h+1}}{u_h} + \frac{v_{k-1} + v_{k+1}}{v_k} = 2\frac{w_{\ell-1} + w_{\ell+1}}{w_\ell},$$

which splits in the following three recurrences

$$\frac{u_{h-1} + u_{h+1}}{u_h} = 2\lambda, \quad \frac{v_{k-1} + v_{k+1}}{v_k} = 2\mu, \quad \frac{w_{\ell-1} + w_{\ell+1}}{w_\ell} = \lambda + \mu.$$

We find successively, as in case of vibrating string,

$$(u_h)_m = A_m \sin\frac{\pi m}{M}h, \quad m, h = \overline{1, M-1},$$

$$(v_k)_n = B_n \sin\frac{\pi n}{N}k, \quad n, k = \overline{1, N-1},$$

$$(w_\ell)_{mn} = C_{mn} \cos(\theta_{mn}\ell) + D_{mn} \sin(\theta_{mn}\ell),$$

where

$$\cos\theta_{mn} = \cos\left(\frac{m}{M} + \frac{n}{N}\right)\frac{\pi}{2} \cos\left(\frac{m}{M} - \frac{n}{N}\right)\frac{\pi}{2}$$

and finally, with new coefficients,

$$u_{hk\ell} = \sum_{m,n=0}^{M-1,N-1} (A_{mn}\cos(\theta_{mn}\ell) + B_{mn}\sin(\theta_{mn}\ell))\sin\frac{\pi mh}{M}\sin\frac{\pi nk}{N}.$$

The coefficients A_{mn} and B_{mn} will be determined using the values for u_{hk0} and u_{hk1} deduced, as above, from the initial conditions.

Remark 4.17: Similar results we have for the linear PDEs:

(i) Spherical wave equation,

$$u_{tt} - a^2(u_{xx} + u_{yy} + u_{zz}) = 0;$$

(ii) Homogeneous heat PDEs,

$$u_t - a^2 u_{xx} = 0,$$

$$u_t - a^2(u_{xx} + u_{yy}) = 0,$$

$$u_t - a^2(u_{xx} + u_{yy} + u_{zz}) = 0.$$

All of these are homogeneous PDEs with constant coefficients.

Solutions of the Korteweg–De Vries PDE, Navier–Stokes PDE, Black–Scholes PDE etc cannot be studied directly by the above method.

4.7 Linearization of Nonlinear ODEs and PDEs

The nonlinear ODEs and PDEs systems modeling real-world phenomena can be very complicated, if not impossible, to be solved explicitly. To be able to analyze these systems we will linearize them.

Our goal is to begin with a nonlinear system, linearize it around a solution, then use linear ODE techniques to understand the approximate behavior of solutions to the linearized system, and finally apply our understanding of the behavior to the nonlinear system.

Linearization of ODE or PDE is required for certain types of analysis such as stability analysis, solution with a Laplace transform, finding small oscillations of system around a solution etc.

For ODEs or PDEs, powerful computational tools such as Maple can be used to perform linearization. Some details about ODE or PDE linearization can be found in the papers [1, 74].

The technique of linearizing nonlinear differential and partial differential equations by variations of solutions is less known in the student world, although it leads immediately to the result.

Problem 4.7.1: **(Linearization of nonlinear Euler–Lagrange ODE)** Linearize a non-linear Euler–Lagrange ODEs system

$$\frac{\partial L}{\partial x^i} - \frac{\partial^2 L}{\partial \dot{x}^i \partial t} - \frac{\partial^2 L}{\partial \dot{x}^i \partial x^j} \dot{x}^j(t) - \frac{\partial^2 L}{\partial \dot{x}^i \partial \dot{x}^j} \ddot{x}^j(t) = 0,$$

$i = \overline{1,n}$, $x(t_0) = x_0$, $x(t_1) = x_1$, around a solution $x(t)$, $t \in [t_0, t_1]$.

Solution. Suppose that the differentiable variation $x(t, \varepsilon)$ satisfies the given system and the end conditions. Partially deriving with respect to ε, i.e., using the operator $D_\varepsilon = \frac{\partial}{\partial x^k} x_\varepsilon^k + \frac{\partial}{\partial \dot{x}^k} \dot{x}_\varepsilon^k$, and denoting $\partial_\varepsilon x|_{\varepsilon=0}(t) = \xi(t)$, we find the linearized ODE (the equation in variations)

$$D\left[\frac{\partial L}{\partial x^i} - \frac{\partial^2 L}{\partial \dot{x}^i \partial t}\right]\bigg|_{\varepsilon=0} - D\left[\frac{\partial^2 L}{\partial \dot{x}^i \partial x^j}\right]\bigg|_{\varepsilon=0} \dot{x}^j(t) - \frac{\partial^2 L}{\partial \dot{x}^i \partial x^j}\bigg|_{\varepsilon=0} \dot{\xi}^j(t)$$

$$- D\left[\frac{\partial^2 L}{\partial \dot{x}^i \partial \dot{x}^j}\right]\bigg|_{\varepsilon=0} \ddot{x}^j(t) - \frac{\partial^2 L}{\partial \dot{x}^i \partial \dot{x}^j}\bigg|_{\varepsilon=0} \ddot{\xi}^j(t) = 0,$$

with unknown $\xi(t)$, satisfying $\xi(t_0) = 0, \xi(t_1) = 0$.

Problem 4.7.2: **(Blausius ODE)** Linearize the ODE

$$\frac{d^3 y}{dx^3}(x) + y(x)\frac{d^2 y}{dx^2}(x) = 0$$

around a solution $y(x)$.

Solution. Let $y(x)$ be a solution of Blausius ODE. Suppose that the differentiable variation $y(x, \varepsilon)$ satisfies

$$\frac{d^3 y}{dx^3}(x, \varepsilon) + y(x, \varepsilon)\frac{d^2 y}{dx^2}(x, \varepsilon) = 0$$

and the condition $y(x, 0) = y(x)$. Partially deriving with respect to ε and denoting

$$\partial_\varepsilon y|_{\varepsilon=0}(x) = \xi(x),$$

we find the linearized (the equation in variations)

$$\frac{d^3 \xi}{dx^3}(x) + y(x)\frac{d^2 \xi}{dx^2}(x) + \frac{d^2 y}{dx^2}(x)\xi(x) = 0,$$

with unknown $\xi(x)$.

Problem 4.7.3: **(Langmuir–Blodgett ODE)** Linearize the ODE

$$\sqrt{y(x)}\,\frac{d^2 y}{dx^2}(x) = e^x$$

around a solution $y(x)$.

Solution. Denote by $y(x)$ a fixed solution of Langmuir–Blodgett ODE. Suppose a differentiable variation $y(x, \varepsilon)$ verifies the ODE

$$\sqrt{y(x, \varepsilon)} \, \frac{d^2 y}{dx^2}(x, \varepsilon) = e^x, \quad y(x, 0) = y(x).$$

Let us consider $\partial_\varepsilon y|_{\varepsilon=0}(x) = \xi(x)$. Then

$$\frac{1}{2} y^{-\frac{1}{2}}(x) \, \frac{d^2 y}{dx^2}(x) \, \xi(x) + \sqrt{y(x)} \, \frac{d^2 \xi}{d\xi^2}(x) = 0.$$

Problem 4.7.4: (Panlevé I transcendent ODE) Linearize the ODE

$$\frac{d^2 y}{dt^2} = 6y^2 + t$$

around a solution $y(t)$.

Problem 4.7.5: (Thomas–Fermi ODE) Linearize the ODE

$$\frac{d^2 y}{dx^2} = \frac{1}{\sqrt{x}} \, y^{3/2}$$

around a solution $y(x)$.

Problem 4.7.6: (Jacobi fields) Let $g_{ij}(x)$ be a Riemannian metric on n-dimensional manifold M, and let Γ^i_{jk} be the Christoffel symbols. Linearize the geodesics ODEs system

$$\ddot{x}^i(t) + \Gamma^i_{jk}(x(t))\dot{x}^j(t)\dot{x}^k(t) = 0$$

around a solution $x^i(t)$.

Hint. The Jacobi field is $\partial_\varepsilon x|_{\varepsilon=0}(t) = \xi(t)$. The linear system is $\frac{D^2}{dt^2}\xi(t) + R(\xi(t), \dot{x}(t))\dot{x}(t) = 0$, where D denotes the covariant derivative with respect to the Levi-Civita connection, and $R(\cdot, \cdot)\cdot$ is the Riemannian curvature tensor. The vector fields $\dot{x}(t)$ and $t\dot{x}(t)$ are Jacobi fields.

Problem 4.7.7: (Linearization of nonlinear Euler–Lagrange PDE) Linearize a nonlinear Euler–Lagrange PDEs system

$$\frac{\partial L}{\partial x^i} - \frac{\partial^2 L}{\partial x^i_\gamma \partial t^\gamma} - \frac{\partial^2 L}{\partial x^i_\gamma \partial x^j} x^j_\gamma(t) - \frac{\partial^2 L}{\partial x^i_\gamma \partial x^j_\lambda} x^j_{\gamma\lambda}(t) = 0, \quad x|_{\partial\Omega} = f$$

around a solution $x(t), t \in \Omega$.

Solution. Suppose that the differentiable variation $x(t, \varepsilon)$ satisfies the given system and the boundary conditions. Partially deriving with respect to ε, i.e., using the operator $D_\varepsilon = \frac{\partial}{\partial x^k} x^k_\varepsilon + \frac{\partial}{\partial x^k_\mu} x^k_{\mu\varepsilon}$, and denoting $\partial_\varepsilon x|_{\varepsilon=0}(t) = \xi(t)$, we find the linearized PDE (the equation in variations)

$$D\left[\frac{\partial L}{\partial x^i} - \frac{\partial^2 L}{\partial x^i_\gamma \partial t^\gamma}\right]\Big|_{\varepsilon=0} - D\frac{\partial^2 L}{\partial x^i_\gamma \partial x^j}\Big|_{\varepsilon=0} x^j_\gamma(t) - \frac{\partial^2 L}{\partial x^i_\gamma \partial x^j}\Big|_{\varepsilon=0} \xi^j_\gamma(t)$$

$$- D\frac{\partial^2 L}{\partial x^i_\gamma \partial x^j_\lambda}\Big|_{\varepsilon=0} x^j_{\gamma\lambda}(t) - \frac{\partial^2 L}{\partial x^i_\gamma \partial x^j_\lambda}\Big|_{\varepsilon=0} \xi^j_{\gamma\lambda}(t) = 0,$$

with unknown $\xi(t)$, satisfying $\xi|_{\partial\Omega} = 0$.

Problem 4.7.8: **(Minimal surface PDE)** Linearize the minimal surfaces PDE

$$(1 + f_y^2)f_{xx} - 2f_x f_y f_{xy} + (1 + f_x^2)f_{yy} = 0.$$

Solution. This is one of the most studied nonlinear equations with partial derivatives. The type of this equation is elliptical, because its linearization is elliptical. Indeed, if the differentiable variation $f(x, y, \varepsilon)$ satisfies this equation and $f(x, y, 0) = f(x, y)$, partially deriving in relation to ε and denoting $\partial_\varepsilon f|_{\varepsilon=0} = \xi$, we find the linearization (equation in variations)

$$2f_y f_{xx}\xi_y + (1 + f_y^2)\xi_{xx} - 2f_y f_{xy}\xi_x - 2f_x f_{xy}\xi_y$$

$$-2f_x f_y \xi_{xy} + 2f_x f_{yy}\xi_x + (1 + f_x^2)\xi_{yy} = 0,$$

which turns out to be elliptical.

Problem 4.7.9: Linearize the nonlinear PDE

$$\frac{\partial u}{\partial t}(x, t) = u^2(x, t)\frac{\partial^2 u}{\partial x^2}(x, t),$$

around a solution $u(x, t)$.

Solution. Accepting that the differentiable variation $u(x, t, \varepsilon)$, of the function $u(x, t, 0) = u(x, t)$, satisfies this ODE, partially deriving with respect to ε and denoting $\partial_\varepsilon u|_{\varepsilon=0} = \xi$, we find the linearized equation (the equation in variations)

$$\frac{\partial \xi}{\partial t}(x, t) = u^2(x, t)\frac{\partial^2 \xi}{\partial x^2}(x, t) + 2u(x, t)\frac{\partial^2 u}{\partial x^2}(x, t)\xi(x, t).$$

Problem 4.7.10: **(Bateman–Burgers PDE)** Linearize the PDE

$$u_t + uu_x = c u_{xx}$$

around a solution $u(x, t)$.

Problem 4.7.11: **(Bookmaster PDE)** Linearize the PDE

$$u_t = (u^4)_{xx} + (u^3)_x$$

around a solution $u(x, t)$.

Problem 4.7.12: **(Dym PDE)** Linearize the PDE

$$u_t = u^3 u_{xxx}$$

around a solution $u(x, t)$.

Problem 4.7.13: Linearize the nonlinear PDE

$$u_t = u(1 - u) + u_{x^2}, \ 0 < x < L, \ t \in [0, \infty),$$

with boundary conditions $u_x(0, t) = 0$, $u_x(L, t) = 0$.

Solution. On the right-hand member, we have the term $u(1 - u)$ that makes a nonlinear PDE.

The PDE has two uniform equilibrium solutions, $u(x, t) = 0$ and $u(x, t) = 1$ (these are easy enough to find: equilibrium means $u_t = 0$, and uniform means $u_{x^2} = 0$, so the uniform equilibrium solutions are the solutions to $u(1 - u) = 0$). The constant solutions $u = 0$ and $u = 1$ also satisfy the boundary conditions, so these are the uniform equilibrium solutions to the PDE.

Now suppose that the initial conditions to the problem are such that $u(x, t)$ is initially close to $u(x, t) = 0$. That is, we consider $u(x, t) = 0 + \varepsilon w(x, t)$, where ε is "small". If we make this substitution into the PDE, we obtain

$$w_t = w - \varepsilon w^2 + w_{x^2}.$$

Now take the limit $\varepsilon \to 0$. We obtain the linearization of PDE at $u = 0$, namely $w_t = w + w_{x^2}$.

4.8 Von Neumann Analysis of Linearized Discrete Tzitzeica PDE

This section applies the von Neumann stability analysis to a discrete Tzitzeica PDE (see [70]).

We need a C^2 function $(u, v) \mapsto \omega(u, v)$. The Tzitzeica hyperbolic PDE

$$\omega_{uv} = e^\omega - e^{-2\omega}$$

is the Euler–Lagrange PDE provided by the first-order Tzitzeica Lagrangian

$$L_T = \frac{1}{2} \omega_u \omega_v + e^\omega - \frac{1}{2} e^{-2\omega}.$$

We attach the moments

$$p = \frac{\partial L_T}{\partial \omega_u} = \frac{1}{2} \omega_v, \ q = \frac{\partial L_T}{\partial \omega_u} = \frac{1}{2} \omega_u$$

and the Hamiltonian PDEs

$$\frac{\partial \omega}{\partial u} = 2q, \ \frac{\partial \omega}{\partial v} = 2p, \ \frac{\partial p}{\partial u} + \frac{\partial q}{\partial v} = e^\omega - e^{-2\omega}.$$

The discretization of a two-parameter Lagrangian $L(\omega(u,v), \omega_u, \omega_v)$ can be performed by using the centroid rule which consists in the substitution of the point $\omega(u,v)$ with the fraction $\frac{1}{3}(\omega_{k\ell} + \omega_{k+1\ell} + \omega_{k\ell+1})$ and of the partial velocities ω_u, ω_v by the fractions $\frac{\omega_{k+1\ell} - \omega_{k\ell}}{k_1}$, $\frac{\omega_{k\ell+1} - \omega_{k\ell}}{k_2}$, where (k_1, k_2) is a fixed step. It follows the *variational integrator*

$$\sum_{\xi} \frac{\partial L}{\partial \omega_{k\ell}}(\xi) = 0,$$

where ξ runs over three points,

$(\omega_{k\ell}, \omega_{k+1\ell}, \omega_{k\ell+1})$, starting point,

$(\omega_{k-1\ell}, \omega_{k\ell}, \omega_{k-1\ell+1})$, left shift point,

$(\omega_{k\ell-1}, \omega_{k+1\ell-1}, \omega_{k\ell})$, right shift point,

and $k = \overline{1, N-1}$, $\ell = \overline{1, M-1}$.

We introduce the discrete momenta via a discrete Legendre transformation

$$p^{k\ell} = \frac{\partial L}{\partial \omega_{k\ell}}(\omega_{k\ell}, \omega_{k+1\ell}, \omega_{k\ell+1}).$$

Then the variational integrator becomes a linear initial value problem with constant coefficients

$$p^{k\ell} + p^{k-1\ell} + p^{k\ell-1} = 0,$$

called *dual variational integrator equation*. The monomial solutions of dual variational integrator are of the form $p^{k\ell} = \beta r_1^k r_2^\ell$. It follows the characteristic equation $r_1 + r_2 + r_1 r_2 = 0$, i.e., r_1 and r_2 are solutions of second-order equation $ar^2 + br + b = 0$. We consider the case $a > 0$ and denote $\Delta = b(b - 4a)$. If $b < 0$ or $b > 4a$, then $\Delta > 0$ and the equation has real solutions r_1, r_2. A solution of the dual integrator is $p^{k\ell} = \beta r_1^k r_2^\ell$. If $0 < b < 4a$, then $\Delta < 0$. In this case, the roots are $r_1 = \rho e^{j\alpha}$, $r_2 = \rho e^{-j\alpha}$ and the solution of the dual integrator is $p^{k\ell} = \beta \rho^{k+\ell} e^{j(k-\ell)\alpha}$.

The discrete Tzitzeica Lagrangian

$$L_d = \frac{1}{2} \frac{\omega_{k+1\ell} - \omega_{k\ell}}{k_1} \frac{\omega_{k\ell+1} - \omega_{k\ell}}{k_2} + e^{(\omega_{k\ell} + \omega_{k+1\ell} + \omega_{k\ell+1})/3}$$

$$+ \frac{1}{2} e^{-2(\omega_{k\ell} + \omega_{k+1\ell} + \omega_{k\ell+1})/3}$$

determines the discrete Tzitzeica equation (discrete Euler–Lagrange equation)

$$\frac{1}{k_1 k_2}(\omega_{k-1\ell} + \omega_{k\ell-1}) - \frac{1}{2k_1 k_2}(\omega_{k\ell+1} + \omega_{k+1\ell} + \omega_{k-1\ell+1} + \omega_{k+1\ell-1})$$

$$+ \frac{1}{3} e^{(\omega_{k\ell} + \omega_{k+1\ell} + \omega_{k\ell+1})/3} + \frac{1}{3} e^{(\omega_{k-1\ell} + \omega_{k\ell} + \omega_{k-1\ell+1})/3}$$

$$+ \frac{1}{3} e^{(\omega_{k\ell-1} + \omega_{k+1\ell-1} + \omega_{k\ell})/3} - \frac{1}{3} e^{-2(\omega_{k\ell} + \omega_{k+1\ell} + \omega_{k\ell+1})/3}$$

$$- \frac{1}{3} e^{-2(\omega_{k-1\ell} + \omega_{k\ell} + \omega_{k-1\ell+1})/3} - \frac{1}{3} e^{-2(\omega_{k\ell-1} + \omega_{k+1\ell-1} + \omega_{k\ell})/3} = 0.$$

This is a second-order nonlinear implicit finite difference equation. The *singularity set* with respect to the variable $u = \omega_{k\ell+1}$ is defined by the equation

$$e^{(\omega_{k\ell} + \omega_{k+1\ell} + \omega_{k\ell+1})/3} + 2 e^{-2(\omega_{k\ell} + \omega_{k+1\ell} + \omega_{k\ell+1})/3} = \frac{9}{2k_1 k_2},$$

(vanishing of the coefficient of the derivative $\frac{d}{du}$). The singularity set with respect to u is given by positive solutions of third-degree algebraic equation

$$Y^3 - \frac{9}{2k_1 k_2} Y^2 + 2 = 0, \ Y = e^{(\omega_{k\ell} + \omega_{k+1\ell} + u)/3}.$$

4.8.1 Von Neumann Analysis of Dual Variational Integrator Equation

To verify the stability of the dual variational equation

$$p^{k\ell} + p^{k-1\ell} + p^{k\ell-1} = 0,$$

we pass to the frequency domain, accepting that u is a spatial coordinate and v is a temporal coordinate. We need a $1D$ discrete spatial Fourier transform which can be obtained via the substitutions (see [42])

$$p^{k\ell} \mapsto P^\ell(\alpha) e^{j\alpha kh},$$

where α denotes the *radian wave scalar*. We find a second-order linear difference equation (digital filter)

$$P^\ell + P^\ell e^{-j\alpha h} + P^{\ell-1} = 0$$

that needs its stability checked. For this purpose we introduce the z-transform $E(z, \alpha)$ and we must impose that the poles of the recursion do not lie outside the unit circle in the z-plane. To simplify, we accept the initial condition $P^0 = 0$. We obtain the homogeneous linear equation

$$(1 + e^{-j\alpha h} + z^{-1})E = 0.$$

The pole $z = \frac{1}{1+e^{-j\alpha h}}$ satisfies the condition

$$\|z\| = \frac{1}{2 \left| \cos \frac{\alpha h}{2} \right|} \leq 1$$

if and only if $\left| \cos \frac{\alpha h}{2} \right| \geq \frac{1}{2}$. This condition ensures that our scheme is marginally stable, and over the stability region, we have a relation in terms of the grid spacing h and the wave number α.

4.8.2 Von Neumann Analysis of Linearized Discrete Tzitzeica Equation

Stability, in general, can be difficult to investigate, especially when equation under consideration is nonlinear. Unfortunately, von Neumann stability is necessary and sufficient for stability in the sense of Lax–Richtmyer (as used in the Lax equivalence theorem) only in certain cases: the PDE and implicitly the finite difference scheme must be linear; the PDE must be constant-coefficient with periodic boundary conditions and have at least two independent variables; and the scheme must use no more than two time levels. It is necessary in a much wider variety of cases, however, and due to its relative simplicity it is often used in place of a more detailed stability analysis as a good guess at the restrictions (if any) on the step sizes used in the scheme.

The linearization of discrete Tzitzeica equation is

$$\frac{1}{k_1 k_2}(\omega_{k-1\ell} - \omega_{k\ell-1}) - \frac{1}{2k_1 k_2}(\omega_{k\ell+1} + \omega_{k+1\ell} + \omega_{k-1\ell+1} + \omega_{k+1\ell-1})$$

$$+ \frac{1}{3}(\omega_{k+1\ell} + \omega_{k\ell+1} + \omega_{k-1\ell} + \omega_{k\ell-1}) + \frac{1}{3}(\omega_{k-1\ell+1} + \omega_{k+1\ell-1}) = 0.$$

Let us find a solution via eigenvalues and a Maple program. We transcribe the previous Tzitzeica system in the form

$$\omega(k-1,l) * (1/(k1 * k2) + 1/3) + \omega(k,l-1) * (-1/(k1 * k2) + 1/3)$$

$$+ \omega(k,l+1) * (-1/(2 * k1 * k2) + 1/3) + \omega(k+1,l) * (-1/(2 * k1 * k2) + 1/3)$$

$$+ \omega(k-1,l+1) * (-1/(2 * k1 * k2) + 1/3) = 0.$$

We are trying solutions of the monomial form $\omega(k,l) = C r_1^k r_2^l$. Denoting $r2 = x$, and setting

> k1 := 2; k2 := 4; r1 := .8

we find the equation

> eq1 :=

$$x^2 * (r1 + 1) * (-1/(2 * k1 * k2) + 1/3) + x * (1/(k1 * k2) + 1/3$$

$$+ r1^2 * (-1/(2 * k1 * k2) + 1/3)) + r1 * (-1/(k1 * k2) + 1/3)$$

$$+ r1^2 * (-1/(2 * k1 * k2) + 1/3) = 0,$$

$$eq1 := .4874999999 * x^2 + .6316666666 * x + .3399999999 = 0.$$

> solve(eq1);

$$-.6478632479 + .5269811281 * I, -.6478632479 - .5269811281 * I$$

> a := .6478632479; b := .5269811281; ρ := $\sqrt{a^2 + b^2}$;
α := arccos(a/ρ)
> with(plots);
plot3d($r1^k * \rho^l * \cos(l * \alpha), k = 0..18, l = 0..19$); see Fig. 4.8
plot3d($r1^k * \rho^l * \sin(l * \alpha), k = 0..18, l = 0..19$); see Fig. 4.9

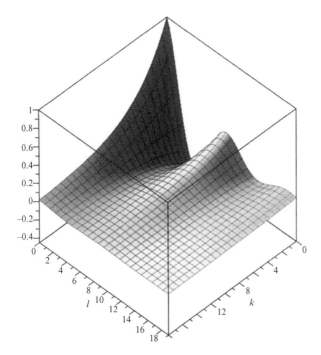

Figure 4.8   Discrete Tzitzeica surface 1.

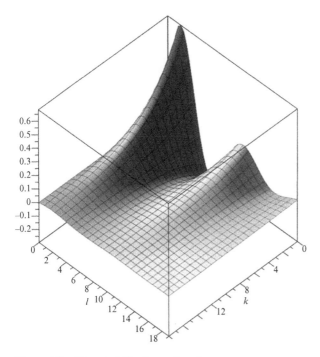

Figure 4.9   Discrete Tzitzeica surface 2.

To verify the stability of previous Tzitzeica finite difference scheme, we pass to the frequency domain, through what is called von Neumann analysis. For that we accept that:

(i) u is a spatial coordinate and v is a temporal coordinate;
(ii) grid spacing in u is uniform, i.e., $h = k_1$ is constant, and we have an unbounded domain $\mathbb{R}$.

We denote by $\tau = k_2$ the "time" step regarding v and we introduce the constant level sets $\dfrac{3}{2k_1 k_2} = \rho$.

We need a $1D$ discrete spatial Fourier transform which can be obtained via the substitutions $\omega_{k\ell} \mapsto \Omega_\ell(\alpha)e^{j\alpha kh}$, where α denotes the radian wave scalar. We find a second-order linear difference equation (digital filter)

$$\rho(2\Omega_\ell e^{-j\alpha h} + 2\Omega_{\ell-1} - \Omega_{\ell+1}) - \rho(\Omega_\ell e^{j\alpha h} + \Omega_{\ell+1}e^{-j\alpha h} - \Omega_{\ell-1}e^{j\alpha h})$$
$$+ 3\Omega_\ell + \Omega_{\ell+1} + \Omega_{\ell-1} + \Omega_\ell e^{j\alpha h} + \Omega_\ell e^{-j\alpha h} + \Omega_{\ell+1}e^{-j\alpha h} + \Omega_{\ell-1}e^{j\alpha h} = 0$$

that needs its stability checked. For this purpose we introduce the z-transform $F(z, \alpha)$ and we must impose the requirement that the poles of the recursion do not lie outside the unit circle in the z-plane. To simplify, we accept the initial condition $\Omega_0 = 0$. One obtains the homogeneous linear equation

$$(1 - \rho)z(1 + e^{-j\alpha h})F + z^{-1}(1 + 2\rho + (1 - \rho)e^{j\alpha h})F$$
$$+ ((1 + 2\rho)e^{-j\alpha h} + (1 - \rho)e^{j\alpha h} + 3)F = 0.$$

The poles are the roots of the characteristic equation

$$(1 - \rho)(1 + e^{-j\alpha h})z^2 + (1 + 2\rho + (1 - \rho)e^{j\alpha h})$$
$$+ ((1 + 2\rho)e^{-j\alpha h} + (1 - \rho)e^{j\alpha h} + 3)z = 0,$$

with the unknown z. Explicitly, we have a second-order algebraic equation

$$a_2 z^2 + a_1 z + a_0 = 0,$$

with the coefficients

$$a_2 = (1 - \rho)(1 + e^{-j\alpha h}), \quad a_1 = (1 + 2\rho)e^{-j\alpha h} + (1 - \rho)e^{j\alpha h} + 3,$$
$$a_0 = 1 + 2\rho + (1 - \rho)e^{j\alpha h}.$$

Since the solutions are

$$z_{1,2} = \frac{-a_1 \pm \sqrt{\Delta}}{2a_2}, \quad \Delta = a_1^2 - 4a_2 a_0,$$

the condition $\|z_{1,2}\| \le 1$ is equivalent to $\eta_{1,2} \ge 0$, where

$$\eta_{1,2} = 4(1 - \rho)\sqrt{2 - 2\cos(\alpha h)} - |-a_1 \pm \sqrt{\Delta}|.$$

Case 1. For $\Im m(a_2) = 0$ all the coefficients of the equation are real numbers. We find two results:

(i) $h = \frac{\pi}{\alpha}$, when $\cos(\alpha h) = -1$. Then we have $a_2 = 2(1 - \rho), a_1 = 1 - \rho, a_0 = 3\rho$ and we find that the double inequality $\frac{1}{13} < \rho < \frac{5}{18}$ ensures the marginal stability to our scheme.

(ii) $h = \frac{2\pi}{\alpha}$, when $\cos(\alpha h) = 1$. Then we have $a_2 = 0, a_1 = 5 + \rho, a_0 = 2 + \rho$. The condition $\|z_{1,2}\| \leq 1$ is satisfied for any $0 < \rho < 1$.

Over the stability region, the previous constraints give us a time step τ, in terms of the grid spacing h, and implicitly in terms of the wave number α.

Case 2. For $\Im m(a_2) \neq 0$, we find that the stability condition $\|z_{1,2}\| \leq 1$ is satisfied also if $\cos(\alpha h) \neq 0$.

Positive values of η_1 and η_2 satisfy the constraints $\|z_{1,2}\| \leq 1$ on the time step τ, the grid spacing h and spatial wave number α, describing the stability region where our previous scheme is marginally stable.

4.9 Maple Application Topics

It has long been thought that ODEs and PDEs, which are continuous, are more funda-mental than difference equations, which are discrete. This opinion relies on the belief that difference equations necessarily derive from ODEs and PDEs. In our opinion, this is not always the case. There may be discrete equations that do not come from ODEs or PDEs. Indeed, some phenomena of nature are most adequately described not in terms of contin-uous but in terms of discontinuous steps (for instance many models in biology, economy, neural processes, decision processes etc).

Example 4.2: (i) *Mandelbrot iteration,*

$$x_{n+1} = x_n^2 - y_n^2 + a_0, \quad y_{n+1} = 2x_n y_n + b_0.$$

(ii) *Benford recurrence,*

$$a_{n+1} = f(n)a_n + g(n)a_{n-1},$$

for certain families of functions f and g.

(iii) *The general case of a second-order constant-coefficient, linear finite difference, homoge-neous equation is*

$$ay_{n+1} + by_n + cy_{n-1} = 0.$$

It is clear that this equation does not always come from the discretization of an ODE. To find solutions, we use the characteristic equation

$$ar^2 + br + c = 0,$$

which is obtained by exploring a simple trial solution of the form $y_n = r^n$, where r is to be determined. We follow the two steps:

(i) if the roots r_1, r_2 are distinct (real or complex), the general solution is given by a linear combination of r_1^n, r_2^n (superposition of solutions), i.e., $y_n = c_1 r_1^n + c_2 r_2^n$;

(ii) if the roots are not distinct, i.e., if $r_1 = r_2$, then we have two independent solutions r_1^n and nr_1^n, and the general solution is given by $y_n = (c_1 + c_2 n) r_1^n$.

In both cases, c_1, c_2 are arbitrary constants of the solution. They can be fixed by specifying initial data, such as the values of y_0, y_1.

Problem 4.9.1: Solve the recurrence

$$2x(n) - x(n-1) = 2^n, \ x(0) = 3.$$

Solution. The general solution of the homogenous equation $2x(n) - x(n-1) = 0$ is $x(n) = c\left(\frac{1}{2}\right)^n$. To find a particular solution of the inhomogeneous problem we try an exponential function $x(n) = d \, 2^n$, with a constant d to be determined. Plugging into the equation we get $d = \frac{2}{3}$. So the general solution is $x(n) = c\left(\frac{1}{2}\right)^n + \frac{2}{3} 2^n$. Adding the initial condition, we find $x(0) = 3 = c + \frac{2}{3}$, and hence $x(n) = \frac{7}{3}\left(\frac{1}{2}\right)^n + \frac{2}{3} 2^n$.

Problem 4.9.2: (Falling object) The discrete Lagrangian is

$$L_d(z_k, z_{k+1}) = m\left[\frac{1}{2}\left(\frac{z_{k+1} - z_k}{h}\right)^2 - g\left(\frac{z_{k+1} + z_k}{2}\right)\right].$$

Solve the discrete Euler–Lagrange equation:

$$\frac{z_{k+1} - 2z_k + z_{k-1}}{h^2} = -g.$$

Hint. We can use Maple command
> rsolve($z(n+1) - 2 * z(n) + z(n-1) = c, z(n)$);
to produce the solution

$$z(n) = c_1 + c_2 - n^2 \, gh^2/2.$$

Problem 4.9.3: Each of the fundamental laws of physics can be written in terms of an action principle. This includes electromagnetism, general relativity, the standard model of particle physics, and attempts to go beyond the known laws of physics such as string theory. For example, (nearly) everything we know about the universe is captured in the Lagrangian

$$L = \sqrt{g}\left(R - \frac{1}{2}F_{\mu\nu}F^{\mu\nu} + \bar{\psi}\,\slashed{D}\psi\right),$$

where the terms carry the names of Einstein, Maxwell (or Yang and Mills) and Dirac respectively, and describe gravity, the forces of nature (electromagnetism and the nuclear forces) and the dynamics of particles like electrons and quarks.

Write the discrete version of L and realize a numerical study of the motion.

Problem 4.9.4: (Rotating coordinate systems) Consider a free particle with Lagrangian given by $L = \frac{1}{2}m\dot{\mathbf{r}}^2$, where $\mathbf{r} = (x, y, z)$. Now measure the motion of the particle with respect to a coordinate system which is rotating with angular velocity $\omega = (0, 0, \omega)$ around the z-axis. If $\mathbf{r}' = (x', y', z')$ are the coordinates in the rotating system, we have the relationship

$$x' = x \cos \omega t + y \sin \omega t, \; y' = y \cos \omega t - x \sin \omega t, \; z' = z.$$

Then we can substitute these expressions into the Lagrangian to find L in terms of the rotating coordinates,

$$L = \frac{1}{2}m\left((\dot{x}' - \omega\dot{y}')^2 + (\dot{y}' + \omega\dot{x}')^2 + \dot{z}^2\right).$$

Derive the equations of motion.

Problem 4.9.5: (Fig. 4.10) We have a vertical torus of radius r and mass M, and a pearl of mass m which can slide freely and without friction around the torus. The torus is spinning around a vertical axis at an angular speed $\dot{\phi}$. The pearl has a velocity component $r\dot{\theta}$ because it is sliding around the torus, and a component $r \sin \theta \dot{\phi}$ because the torus is spinning. The resultant speed is the orthogonal sum of these. Find the kinetic energy and the potential energy of the system. Write the Euler–Lagrange ODEs.

Solution. The kinetic energy of the system is the sum of the translational kinetic energy of the pearl and the rotational kinetic energy of the torus

$$T = \frac{1}{2}mr^2 \left(\dot{\theta}^2 + \sin^2 \theta \dot{\phi}^2\right) + \frac{1}{2}\left(\frac{1}{2}Mr^2\right)\dot{\phi}^2.$$

If we refer potential energy to the center of the torus, then $V = mgr \cos \theta$. Denoting $L = T - V$, it follows the Euler–Lagrange ODEs

$$r\left(\ddot{\theta} - \sin \theta \cos \theta \dot{\phi}^2\right) - g \sin \theta = 0, \; m \sin^2 \theta \dot{\phi} + \frac{1}{2}M\dot{\phi} = constant.$$

Figure 4.10 Problem 4.9.5.

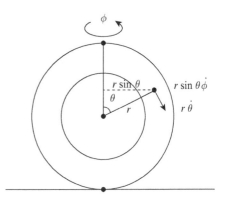

Problem 4.9.6: Using the Euler–Lagrange ODE, find the extremals for the simple integral functional

$$I(y) = \int_a^b \sqrt{y(x)(1 + y'(x)^2)} \, dx.$$

Problem 4.9.7: Reduce the boundary value problem

$$y''(x) - y(x) + x = 0, \quad y(0) = y(1) = 0$$

into a variational problem.

Hint. Multiply both sides of given ODE by δy and integrate over the interval $(0, 1)$. Integrating by parts, we obtain the form $\delta I(y) = 0$. Consequently: extremize $I(y) = \int_0^1 (y'(x)^2 + y(x)^2 - 2xy(x)) \, dx$, $y(0) = y(1) = 0$.

Another hint. Any linear second-order ODE can be written in self-adjoint form $\frac{d}{dx}(p(x)y'(x)) + q(x) = 0$, or $p(x)y''(x) + p'(x)y'(x) + q(x) = 0$. Comparing with the given equation we find $\frac{p}{1} = \frac{p'}{0} = \frac{q}{-y+x}$, from where it is obtained $\frac{d}{dx}(Cy'(x)) - C(y(x) - x) = 0$. Finally, taking $C = 2$, and comparing with a generic Euler–Lagrange equation, we find

$$L(x, y(x), y'(x)) = y'(x)^2 + y(x)^2 - 2xy(x).$$

Problem 4.9.8: Reduce each of the boundary value problem

$$y''(x) + p(x)y(x) + q(x) = 0, \quad y(0) = y(1) = 0$$
$$y''(x) + (2x + 1)y(x) + 1 + x^2 = 0, \quad y(0) = y(1) = 0$$

into a variational problem.

5

Miscellaneous Topics

Motto:
"Houses clustered closely – as so many a pitcher
With wine at their bottoms, thick with being old –
Lie on banks of azure, of the sun's stream – richer,
–From whose muddy waters I have sipped pure gold."
Tudor Arghezi – *Never Yet Had Autumn...*

In this chapter, we present other applications selected from science and engineering (magnetic levitation, sensor problem, particle motion in fields, geometric dynamics, wind theory), formulated using the laws of physics. For one-dimensional problems, the boundary conditions are conditions at the ends and will be specified. For multidimensional problems, boundary conditions are more sophisticated.

Traditional books stop only at necessary conditions, whereas the problems dealt with are sure to be solved and once found the solutions, they are and not others. We also specify some sufficient conditions. For details, see [3, 6, 7, 11, 12, 18, 19, 20, 26, 27, 42, 46, 52, 56, 61], [63]–[68], [69, 71, 83].

5.1 Magnetic Levitation

Magnetic levitation (maglev) or magnetic suspension is a method by which an object is suspended with no support other than magnetic fields. Magnetic force is used to counteract the effects of the gravitational force and any other forces. For details, see [56, 61].

In this section we discuss two methods of generating trajectories for magnetic levitation (for a long time, a subject of Science Fiction). The first refers to a nonlinear model, and the second refers to its linearization.

5.1.1 Electric Subsystem

This problem uses a geometric configuration of electric currents that produces the surrounding magnetic field capable of creating the levitation of a ball. In fact, the geometry of electrical wires is able to create magnetic fields with super-sophisticated properties, a fact

Variational Calculus with Engineering Applications, First Edition. Constantin Udriste and Ionel Tevy.
© 2023 John Wiley & Sons Ltd. Published 2023 by John Wiley & Sons Ltd.

known to geophysicists in Romania (the representative figure being Sabba Stefanescu) and Russia (Mikhail Solomonovich Ioffe, configuration magnetic traps).

The electromagnet has a resistance R_c and an inductance that depends on the position of the ball, usually shaped by the function

$$L(x_b) = L_1 + \frac{L_0}{1 + \frac{x_b}{\alpha}},$$

where L_0, L_1, α are positive constants, and x_b indicates the vertical position of the ball. Instead of $L(x_b)$, we shall use a constant value L_c. The system is also equipped with a resistor R_s connected in series with a coil whose voltage can be measured. Kirchhoff's law is equivalent to the ODE

$$V_c(t) = \dot{\Phi}(t) + (R_c + R_s) I_c(t),$$

where $\Phi(t) = L_c I_c(t)$ is the magnetic flux, and t means time.

5.1.2 Electromechanic Subsystem

The electromagnetic force F_c induced by the current $I_c(t)$, which acts on the ball, has the expression

$$F_c(t) = \frac{K_m I_c^2(t)}{2(x_b(t) + a)^2}, \quad x_b(t) \geq 0,$$

where K_m is the constant of electromagnetic force and a is an experimentally determined constant, which prevents the force from becoming infinite when $x_b = 0$.

The total force acting on the ball is

$$-F_c(t) + F_g(t) = -\frac{K_m I_c^2(t)}{2(x_b(t) + a)^2} + M_b g,$$

where F_g is the force of gravity, M_b is the mass of the ball and g is the gravitational constant. The Newton law becomes the second-order ODE

$$\ddot{x}_b(t) = -\frac{K_m I_c^2(t)}{2M_b(x_b(t) + a)^2} + g.$$

This Newton law comes from the variational principle attached to the Lagrangian $L = \frac{1}{2} M_b \dot{x}_b(t)^2 - U(x_b(t))$, where

$$U(x_b(t)) = -\frac{K_m I_c^2(t)}{2(x_b(t) + a)} + M_b g x_b(t).$$

The experimental data cited in [56] are: $M_b = 0.068\,Kg$, $a = 4.2\,mm$, $K_m = 1.94 \times 10^{-4} Nm^2/A^2$.

5.1.3 State Nonlinear Model

Re-note the position of the ball with $x^1 = x_b$, the ball velocity by $x^2 = \dot{x}_b$, current intensity in the coil with $x^3 = I_c$. Also, we set $R = R_c + R_s$ and input voltage $v = V_c$. It follows the state $x = (x^1, x^2, x^3)$ and the first-order differential system (which describes the evolution)

$$\dot{x}^1(t) = x^2(t), \ \dot{x}^2(t) = -\frac{K_m x^3(t)^2}{2M_b(x^1(t) + a)^2} + g, \ \dot{x}^3(t) = \frac{1}{L_c}(-Rx^3(t) + v(t)).$$

The evolution of the electrical part is much faster than the evolution from the electromechanical part. From the design point of view, we can neglect the rapid dynamics, which is to impose control $u(t) = I_c(t)$ and to study the states $x = (x^1, x^2)$ through the controlled system

$$\dot{x}^1(t) = x^2(t), \ \dot{x}^2(t) = -\frac{K_m u^2(t)}{2M_b(x^1(t) + a)^2} + g. \tag{1}$$

5.1.4 The Linearized Model of States

A static equilibrium point $x_{eq} = (x_{b,eq}, 0)$ is characterized by the fact that the ball is suspended in the air, at the optimal position $x_{b,eq}$ due to a constant electromagnetic force generated by i_{eq}. Hence

$$i_{eq} = \sqrt{\frac{2M_b g}{K_m}} \ (x_{b,eq} + a).$$

Denoting $\xi^1 = x^1 - x_{b,eq}$, $\xi^2 = x^2$, $v = u - i_{eq}$, the linear approximation of the differential system (1) is

$$\dot{\xi}^1(t) = \xi^2(t), \ \dot{\xi}^2(t) = \frac{2g}{x_{b,eq} + a}\xi^1(t) - \frac{\sqrt{2K_m M_b g}}{M_b(x_{b,eq} + a)} v(t).$$

The matrix transcription is obvious. It turns out that this differential system is unstable, but controllable (we leave the reader to verify).

5.2 The Problem of Sensors

Sensor networks are present in many practical and theoretical problems. Applications include: environmental monitoring (traffic vehicles, neighborhood security, forest fires etc), industrial or agricultural diagnoses (factory conditions, soil health etc) and developments on the battlefields (movement of combat troops, movement of enemy troops etc). For details, see [26].

Being given a set S of n sensors localized at the points $s_1, ..., s_n$, the power of collective observation with respect to an arbitrary point p is a function of the form

$$I(S, p) = I(f_1(s_1, p), ..., f_n(s_n, p)),$$

called intensity of sensors field (Fig. 5.1). Theorists have proposed two types of intensities: (i) sum intensity of all sensors field

$$I_t(S, p) = \sum_{i=1}^{n} f_i(s_i, p);$$

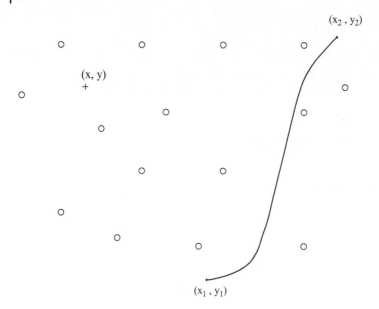

Figure 5.1 Senzors.

(ii) max intensity of sensors field

$$I_{\max}(S, p) = \max_{i}\{f_i(s_i, p) \mid i = \overline{1, n}\}.$$

Exposure of an object moving in a field of sensors in the interval $[t_1, t_2]$, along the path $p(t)$, is defined as a simple integral functional

$$E(p(\cdot)) = \int_{t_1}^{t_2} I(S, p(t)) \left\| \frac{dp}{dt}(t) \right\| dt.$$

Problem 5.2.1: (Exposure problem) Knowing that the path $p(t)$ and the Lagrangian

$$L = I(S, p(t)) \left\| \frac{dp}{dt}(t) \right\|$$

are of class C^2, find minimal exposure.

5.2.1 Simplified Problem

We consider a single sensor located at the point s and denote by $d(s, p)$ distance from the location s of the sensor to an arbitrary location p. We fix the *sensitivity function* as $f(s, p) = \frac{1}{d(s,p)}$ and we are looking for the path $p(t) = (x(t), y(t))$ of minimum exposure between the point $A = (x_a, y_a)$ and the point $B = (x_b, y_b)$, that is, with the boundary conditions $x(t_1) = x_a$, $y(t_1) = y_a$ and $x(t_2) = x_b$, $y(t_2) = y_b$. We must minimize the *exposure functional*

$$E(x(\cdot), y(\cdot)) = \int_{t_1}^{t_2} I(x(t), y(t)) \sqrt{\dot{x}^2(t) + \dot{y}^2(t)} \, dt.$$

For simplification, we pass to polar coordinates (with origin at the point s)

$$x(t) = \rho(t) \cos \theta(t), \quad y(t) = \rho(t) \sin \theta(t).$$

The functional is changed into

$$E(\rho(\cdot), \theta(\cdot)) = \int_{t_1}^{t_2} I(\rho(t), \theta(t)) \sqrt{(\rho\dot{\theta})^2(t) + \dot{\rho}^2(t)} \, dt.$$

Identifying t with θ, denoting the angle $A - senzor - B$ with α and $d(s, A) = a$, $d(s, B) = b$, we find an even more simplified form (see Fig. 5.2):

$$E(\rho(\cdot)) = \int_0^\alpha I(\rho(\theta), \theta) \sqrt{\rho^2(\theta) + \left(\frac{d\rho}{d\theta}\right)^2 (\theta)} \, d\theta, \quad \rho(0) = a, \rho(\alpha) = b.$$

Since $I(\rho(\theta), \theta) = \dfrac{1}{\rho}$, the Lagrangian $L = \dfrac{1}{\rho} \sqrt{\rho^2 + \left(\frac{d\rho}{d\theta}\right)^2}$ does not depend explicitly on θ and consequently the Euler–Lagrange ODE $\dfrac{\partial L}{\partial \rho} - \dfrac{d}{d\theta} \dfrac{\partial L}{\partial \rho_\theta} = 0$ reduces to a first integral

$$\frac{1}{\sqrt{1 + \frac{1}{\rho^2} \left(\frac{d\rho}{d\theta}\right)^2}} = c.$$

This differential equation is also written $\dfrac{d\rho}{d\theta} = -c_1 \rho$ and has the solution $\rho(\theta) = c_2 e^{\pm c_1 \theta}$. Putting the conditions at ends, we find the curve $\rho(\theta) = a e^{\frac{\ln(b/a)}{\alpha} \theta}$. In particular, for $a = b$ we find the circle arc $\rho(\theta) = a$ centered in s and passing by the points A and B. The solution is optimal because the Lagrange-Jacobi criterion is met.

Figure 5.2 Simple exposure.

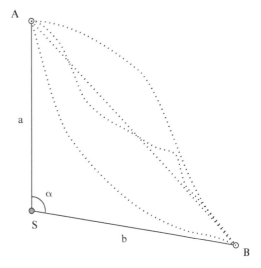

5.2.2 Extending the Simplified Problem of Sensors

Let us introduce a more complicated intensity.

Theorem 5.1: *Suppose the intensity is* $I(s, p) = \frac{1}{d^k(s,p)}$, $k > 0$, $k \neq 1$. *Being given the points A and B, with* $d(s, A) = a$, $d(s, B) = b$ *and* $\alpha \leq \frac{\pi}{k-1}$ *as measure of the angle* $A - senzor - B$, *the path of minimal exposure has the representation*

$$
\rho(\theta) = a \left(\cos(k-1)\theta + (\sin(k-1)\theta) \, \frac{\frac{b^{k-1}}{a^{k-1}} - \cos(k-1)\alpha}{\sin(k-1)\alpha} \right)^{\frac{1}{k-1}}.
$$

5.3 The Movement of a Particle in Non-stationary Gravito-vortex Field

In three-dimensional space $\mathbb{R}^3$, we consider a particle of mass m, marked by the position vector $\vec{r}(x, y, z)$. The particle moves along a curve $x = x(t), y = y(t), z = z(t)$, $t \in \mathbb{R}$, with velocity $\vec{v} = \dot{\vec{r}}(x(t), y(t), z(t))$. The Lagrangian (kinetic energy minus potential energy)

$$
L = \frac{1}{2} m v^2 - m U(\vec{r}), \quad v^2 = \langle \vec{v}, \vec{v} \rangle
$$

produces the Euler–Lagrange ODEs (vectorial form)

$$
\frac{\partial L}{\partial \vec{r}} - \frac{d}{dt} \frac{\partial L}{\partial \vec{v}} = 0.
$$

We introduce: (i) the *absolute velocity* $\vec{v}' = \vec{v} + \vec{\omega} \times \vec{r}$, with respect to a coordinate system K', which consists of velocity $\vec{v}$ in relation to its own coordinate system K, to which is added the rotational velocity $\vec{\omega} \times \vec{r}$ of the system K; (ii) the *vortex vector field* $\vec{\Omega} = 2\vec{\omega}$; (iii) rewriting the absolute velocity $\vec{v}' = \vec{v} + \frac{1}{2}\vec{\Omega} \times \vec{r}$. The vector field $\vec{A} = \frac{1}{2}\vec{\Omega} \times \vec{r}$ is even the *potential vector* of the field ([52]).

The momentum $\vec{P}$ of the particle of mass m, in relation to an arbitrary coordinate system K', is called *generalized momentum* and has the expression

$$
\vec{P} = m\vec{v}' = \vec{p} + m\vec{A}.
$$

It contains the momentum $\vec{p} = m\vec{v}$, in relation to the coordinate system K. On the other hand we know that the momentum $\vec{P}$ is the partial derivative of the Lagrangian with respect to the velocity, i.e., $\vec{P} = \frac{\partial L}{\partial \vec{v}'}$. It turns out that in the coordinate system K', the Lagrangian L is replaced by

$$
L_1 = \frac{1}{2} m v^2 + m <\vec{A}, \vec{v}> -m U(\vec{r}).
$$

The first term is the kinetic energy of a free particle; the last two terms describe the interaction of the particle with the field.

The Euler–Lagrange equations produced by L_1 describe the movement in the gravito-vortex field. To write them, we notice that if we accept $\vec{v} = \nabla\varphi$, then

$$\frac{\partial L_1}{\partial \vec{r}} = \nabla L_1 = m\nabla < \vec{A}, \vec{v} > -m\nabla U$$

$$= m < \vec{v}, \nabla > \vec{A} + m\vec{v} \times \mathrm{rot}\vec{A} - m\nabla U.$$

Furthermore,

$$\frac{d}{dt}\frac{\partial L_1}{\partial \vec{v}} = \frac{d}{dt}\vec{P} = \frac{d\vec{p}}{dt} + m\frac{d\vec{A}}{dt}.$$

Since $\frac{d\vec{A}}{dt} = \frac{\partial \vec{A}}{\partial t} + < \vec{v}, \nabla > \vec{A}$, the equations of motion are

$$\frac{d\vec{p}}{dt} = m\left(-\frac{\partial \vec{A}}{\partial t} - \nabla U\right) + m\vec{v} \times \vec{\Omega}.$$

On the right we recognize a gyroscopic force that consists of the gravitational part $\vec{G} = m\left(-\frac{\partial \vec{A}}{\partial t} - \nabla U\right)$ and gyroscopic part $m\vec{v} \times \vec{\Omega}$ (which generates null mechanical work).

5.4 Geometric Dynamics

Let us explain the meaning of geometric dynamics generated by a flow and a Riemannian metric, respectively an m-flow and two Riemannian metrics.

5.4.1 Single-time Case

Let $X = (X^1, ..., X^n)$ be a C^∞ vector field on the Riemannian manifold $(\mathbb{R}^n, g_{ij})$. This vector field determines the flow

$$\dot{x}^i(t) = X^i(x(t)), \quad i = \overline{1, n}. \tag{1}$$

The classical extension of the differential system (1) is obtained by derivation along a solution, i.e.,

$$\ddot{x}^i(t) = \frac{\partial X^i}{\partial x^j}(x(t))\,\dot{x}^j(t),$$

with sum over the j index. This last differential system has all the solutions of system (1), but it also has other solutions. Coming back we get $\dot{x}^i(t) = X^i(x(t)) + c$.

We intend to change this extension into an Euler–Lagrange prolongation. For this we decompose the right-hand side in the form

$$\ddot{x}^i(t) = \left(\frac{\partial X^i}{\partial x^j} - \frac{\partial X^j}{\partial x^i}\right)(x(t))\,\dot{x}^j(t) + \frac{\partial X^j}{\partial x^i}(x(t))\,\dot{x}^j(t),$$

maneuvering the Jacobian matrix, with sum over the j index. Then, in the last term we replace $\dot{x}^i(t) = X^i(x(t))$. We find the prolongation

$$\ddot{x}^i(t) = \left(\frac{\partial X^i}{\partial x^j} - \frac{\partial X^j}{\partial x^i}\right)(x(t))\,\dot{x}^j(t) + \frac{\partial f}{\partial x^i}(x(t)), \tag{2}$$

where $f(x) = \frac{1}{2}g_{ij}X^i(x)X^j(x)$ is the density of kinetic energy associated with the vector field X. Extension (2) turns out to be an Euler–Lagrange prolongation. Indeed, if we use the *least squares Lagrangian*

$$L = \frac{1}{2}g_{ij}(x)\,(\dot{x}^i - X^i(x))(\dot{x}^j - X^j(x)),$$

then the Euler–Lagrange ODEs

$$\frac{\partial L}{\partial x^i} - \frac{d}{dt}\frac{\partial L}{\partial \dot{x}^i} = 0$$

are just the equations (2). The second-order differential system (2) describes two equivalent notions: (i) a geodesic motion in a gyroscopic field of forces, name given by J. E. Marsden in a review of our book [64]; (ii) geometric dynamics, name given by us (see [63]–[68], [69, 71, 72]). Indeed, on the right-hand side we recognize a gyroscopic force

$$\left(\frac{\partial X^i}{\partial x^j} - \frac{\partial X^j}{\partial x^i}\right)\dot{x}^j$$

(which produces zero mechanical work) and a conservative force ∇f (their sum being a gyroscopic force field).

The least squares Lagrangian L produces the Hamiltonian

$$H = \frac{1}{2}g_{ij}(x)\,(\dot{x}^i - X^i(x))(\dot{x}^j + X^j(x)).$$

5.4.2 The Least Squares Lagrangian in Conditioning Problems

The least squares Lagrangian can be very useful for solving flow-type problems with initial and final conditions.

Example 5.1: *Let us consider a C^1 function $X = X(t,x) : [a,b] \times \mathbb{R} \to \mathbb{R}$ and the following problems:*

Problem 5.4.1: Find $x : [a,b] \to \mathbb{R}$, $x \in C^1$, such that

$$\dot{x}(t) = X(t,x(t)) \text{ with } x(a) = x_1 \text{ and } x(b) = x_2,$$

where $\dot{x}$ is the derivative of x with respect to t.

Generally, such a problem has no solution. Then we intend to find a solution which to be the best in a certain sense:

Problem 5.4.2:   Find a C^1 function $x : [a, b] \to \mathbb{R}$, such that $x(a) = x_1$, $x(b) = x_2$, and

$$\int_a^b [\dot{x}(t) - X(t, x(t))]^2 \, dt = minimum.$$

This means the best approximation in the norm of $L^2([a, b])$.

Solution.   The Euler–Lagrange equation of this variational problem is

$$-(\dot{x} - X(t, x)) \frac{\partial X}{\partial x} - \frac{d}{dt}[(\dot{x} - X(t, x))] = 0.$$

This means either

$$\dot{x} = X(t, x),$$

or

$$\frac{d}{dt}[\ln(\dot{x} - X(t, x))] = -\frac{\partial X}{\partial x}.$$

The last equation is a second-order ODE whose solutions depend on two arbitrary constants and so the conditions $x(a) = x_1$, $x(b) = x_2$ can be fulfilled. The solution is a minimum point because $\dfrac{\partial^2 L}{\partial \dot{x}^2} = 2 > 0$ (Legendre-Jacobi criterium).

Remark 5.1:   If at point b we set the natural boundary condition, $\dfrac{\partial L}{\partial \dot{x}}(b) = 0$, we obtain again the solution of the Cauchy problem with $x(a) = x_1$.

In Problems 5.4.1 and 5.4.2, we take $X(t, x) = tx$. Then the ODE $\dot{x} = tx$ gives us $x(t) = c \, e^{\frac{t^2}{2}}$ and Problem 5.4.1 has solution iff $x_2 = x_1 \, e^{\frac{b^2 - a^2}{2}}$.

In Problem 5.4.2, the Euler–Lagrange equation is $\dfrac{d}{dt}[\ln(\dot{x} - tx)] = -t$, which gives us the general solution

$$x(t) = \left(c_1 + c_2 \int_a^t e^{-s^2} ds\right) e^{\frac{t^2}{2}}.$$

Then we can determine $c_1 = x_1 e^{-a^2/2}$ and a corresponding c_2.

Obviously, if $x_2 = x_1 \, e^{\frac{b^2 - a^2}{2}}$, we find $c_2 = 0$, i.e., the solution of Problem 5.4.1.

Example 5.2:   (**Heat equation**) *For a function of two real variables $u(t, x)$ solve the equation $u_t = u_{xx}$, with boundary condition $u(t, 0) = u(t, L) = 0$, the initial condition $u(0, x) = \varphi(x)$ and the final condition $u(T, x) = \psi(x)$.*

Missing the final condition we find

$$u(t, x) = \sum_{n=1}^{\infty} A_n e^{-\frac{n^2 \pi^2 t}{L^2}} \sin \frac{n \pi x}{L},$$

where $\{A_n\}$ are the Fourier coefficients of the function $\varphi(x)$. As in the above example, adding the final condition, the problem has no solution, generally.

Problem 5.4.3: Find a function of two real variables $u(t, x)$, with boundary condition $u(t, 0) = u(t, L) = 0$, the initial condition $u(0, x) = \varphi(x)$ and the final condition $u(T, x) = \psi(x)$, such that

$$\iint_{\Omega} (u_t(t, x) - u_{xx}(t, x))^2 \, dt dx = minimum,$$

where $\Omega = [0, T] \times [0, L]$. This means the best approximation in the norm of $L^2(\Omega)$.

Solution. The Euler–Lagrange equation becomes

$$\frac{\partial}{\partial t}(u_t - u_{xx}) = \frac{\partial^2}{\partial x^2}(u_t - u_{xx}),$$

which is equivalent to the PDE $w_t - w_{xx} = 0$, where $w = u_t - u_{xx}$. If $K(u) = u_t - u_{xx}$ would be the heat operator, this equation takes the form $K^2(u) = 0$.

Solving the equation by the method of separating the variables, $u(t, x) = T(t)X(x)$, we obtain successively:

$$T''X - T'X'' = T'X'' - TX^{IV},$$

or

$$\frac{T''}{T} - 2\frac{T'}{T}\frac{X''}{X} + \frac{X^{IV}}{X} = 0.$$

This equality takes place for

$$\frac{X^{IV}}{X} = \lambda, \quad \frac{X''}{X} = \mu \quad \text{and} \quad -\frac{T''}{T} = -2\mu\frac{T'}{T} + \lambda.$$

We obtain $\lambda = \mu^2$, then, according to the usual calculations, $\mu = -\frac{n^2\pi^2}{L^2}$, and finally

$$u(t, x) = \sum_{n=1}^{\infty} (A_n t + B_n) e^{-\frac{n^2 \pi^2 t}{L^2}} \sin \frac{n \pi x}{L},$$

where $\{A_n\}$ and $\{B_n\}$ are calculated from the Fourier coefficients of the functions φ and ψ.

Example 5.3: **(Wave equation)** *Similar problems can be proposed for the wave equation $u_{tt} - u_{xx} = 0$, with boundary condition $u(t, 0) = u(t, L) = 0$, the initial conditions $u(0, x) = \varphi_1(x)$, $u_t(0, x) = \varphi_2(x)$ and the final conditions $u(T, x) = \psi_1(x)$, $u_t(T, x) = \psi_2(x)$.*

The least squares Lagrangian in this case is $(u_{tt} - u_{xx})^2$. The Euler–Lagrange equation leads to PDE

$$\frac{\partial^2}{\partial t^2}(u_{tt} - u_{xx}) = \frac{\partial^2}{\partial x^2}(u_{tt} - u_{xx}).$$

The solution of this PDE is expressed by a series containing four sequences of coefficients $\{A_n\}, \{B_n\}, \{C_n\}, \{D_n\}$, which are calculated from the Fourier coefficients of the four functions $\varphi_1(x), \varphi_2(x), \psi_1(x), \psi_2(x)$, i.e.,

$$u(t, x) = \sum_{n=1}^{\infty}\left[(A_n t + B_n)\cos\frac{n^2\pi^2 t}{L^2} + (C_n t + D_n)\sin\frac{n^2\pi^2 t}{L^2}\right]\sin\frac{n\pi x}{L}.$$

5.4.3 Multi-time Case

Let (T, h) and (M, g) be two Riemannian manifolds of dimensions m and n, with $m < n$, in order written here, and called source space and target space. The first manifold (T, h) must be oriented because we use an integral on it, namely the action = integral of a Lagrangian function. The local coordinates are $t = (t^\alpha)$, $\alpha = \overline{1, m}$, respectively $x = (x^i)$, $i = \overline{1, n}$. The metric tensors and the Christoffel symbols have the components $h_{\alpha\beta}, g_{ij}$, respectively $H^\alpha_{\beta\gamma}, G^i_{jk}$.

A d-tensor field (distinguished tensor field) $X^i_\alpha(t, x)$ on $T \times M$ determines the PDEs system $\frac{\partial x^i}{\partial t^\alpha}(t) = X^i_\alpha(t, x(t))$. This PDE system is completely integrable if and only if $\frac{\partial^2 x^i}{\partial t^\alpha \partial t^\beta} = \frac{\partial^2 x^i}{\partial t^\beta \partial t^\alpha}$, i.e.,

$$\frac{\partial X^i_\alpha}{\partial t^\beta} + \frac{\partial X^i_\alpha}{\partial x^j}X^j_\beta = \frac{\partial X^i_\beta}{\partial t^\alpha} + \frac{\partial X^i_\beta}{\partial x^j}X^j_\alpha, \alpha \neq \beta.$$

A Cauchy problem attached to a completely integrable system has unique solution $x = x(t)$ called m-sheet. In other words, the PDEs system determines an m-flow.

Whatever the previous system, we can build the least squares Lagrangian

$$L = \frac{1}{2}h^{\alpha\beta}g_{ij}(x^i_\alpha - X^i_\alpha(t, x))(x^j_\beta - X^j_\beta(t, x))\sqrt{\det(h)},$$

where $x^i_\alpha = \frac{\partial x^i}{\partial t^\alpha}$. The multi-time geometric dynamics is described by Euler–Lagrange PDEs (Poisson PDEs)

$$h^{\alpha\beta}x^i_{\alpha\beta} = g^{ih}h^{\alpha\beta}g_{kj}(\nabla_h X^k_\alpha)X^j_\beta + h^{\alpha\beta}F^i_{j\alpha}x^j_\beta + h^{\alpha\beta}D_\beta X^i_\alpha,$$

where

$$\frac{\delta}{\delta t^\beta}x^i_\alpha := x^i_{\alpha\beta} = \frac{\partial^2 x^i}{\partial t^\alpha \partial t^\beta} - H^\gamma_{\alpha\beta}x^i_\gamma + G^i_{jk}x^j_\alpha x^k_\beta,$$

$$F^i_{j\alpha} = \nabla_j X^i_\alpha - g^{ih}g_{kj}\nabla_h X^k_\alpha,$$

$$\nabla_j X^i_\alpha = \frac{\partial X^i_\alpha}{\partial x^j} + G^i_{jk}X^k_\alpha, \quad D_\beta X^i_\alpha = \frac{\partial X^i_\alpha}{\partial t^\beta} - H^\gamma_{\alpha\beta}X^i_\gamma.$$

5.5 The Movement of Charged Particle in Electromagnetic Field

In this section, we refer to movement under the action of Lorentz force and the movements called unitemporal geometric dynamics, induced by electromagnetic vector fields.

Let U be an isotropic, homogeneous, linear medium domain in the Riemannian manifold $(M = \mathbb{R}^3, \delta_{ij})$. Let ∂_t be the derivation operator with respect to time. The Maxwell PDEs (first-order coupled PDEs)

$$\text{div } D = \rho, \text{ rot } H = J + \partial_t D,$$
$$\text{div } B = 0, \text{ rot } E = -\partial_t B$$

together with the constitutive equations $B = \mu H$, $D = \varepsilon E$ on $U \times \mathbb{R}$, reflect the relationship between ingredients of electromagnetic field:

E	$[V/m]$	electric force field
H	$[A/m]$	magnetic force field
J	$[A/m^2]$	electric current density
ε	$[As/Vm]$	permittivity
μ	$[Vs/Am]$	permeability
B	$[T] = [Vs/m^2]$	magnetic induction (magnetic flux density)
D	$[C/m^2] = [As/m^2]$	electric displacement (electric flux density).

Since $\text{div } B = 0$, the vector field B is without sources, so may be expressed as rot of a potential vector field A, i.e., $B = \text{rot } A$. Then the electric field is $E = -\text{grad } V - \partial_t A$.

It is well known that the motion of a charged particle in an electromagnetic field is described by the ODE system (*Universal Lorentz Law*)

$$m\frac{d^2x}{dt^2} = e\left(E + \frac{dx}{dt} \times B\right), \quad x = (x^1, x^2, x^3) \in U \subset \mathbb{R}^3,$$

where m is the *mass* and e is the particle *charge*. Of course, these Euler–Lagrange equations are produced by the Lorentz Lagrangian

$$L_1 = \frac{1}{2} m\delta_{ij}\frac{dx^i}{dt}\frac{dx^j}{dt} + e\delta_{ij}\frac{dx^i}{dt}A^j - eV.$$

The associated Lorentz Hamiltonian is

$$H_1 = \frac{m}{2}\delta_{ij}\frac{dx^i}{dt}\frac{dx^j}{dt} + eV.$$

A similar mechanical motion was discovered by I. N. Popescu [52], accepting the existence of a *gravito-vortex field* represented by the force $G + \dfrac{dx}{dt} \times \Omega$, where G is the gravitational field and Ω is a vortex determining the gyroscopic force part. Our recent papers [63]–[68], [69, 71, 72] confirm Popescu's point of view through geometric dynamics.

5.5.1 Unitemporal Geometric Dynamics Induced by Vector Potential A

To preserve traditional formulas, we will refer to the field lines of the potential vector "$-A$" using homogeneous dimensional relations. These curves are the solutions of the ODE system

$$m \frac{dx}{dt} = -eA.$$

This system together with the Euclidean metric δ_{ij} produce the least squares Lagrangian

$$L_2 = \frac{1}{2} \delta_{ij} \left(m \frac{dx^i}{dt} + eA^i \right) \left(m \frac{dx^j}{dt} + eA^j \right)$$

$$= \frac{1}{2} \left\| m \frac{dx}{dt} + eA \right\|^2.$$

The Euler–Lagrange ODEs associated to the Lagrangian L_2 are

$$m \frac{d^2 x^i}{dt^2} = e \left(\frac{\partial A^j}{\partial x^i} - \frac{\partial A^i}{\partial x^j} \right) \frac{dx^j}{dt} + \frac{e^2}{m} \frac{\partial f_A}{\partial x^i} - e \partial_t A$$

(sum over j), where

$$f_A = \frac{1}{2} \delta_{ij} A^i A^j$$

is the *energy density* produced by the vector field A. Equivalently, it appears a *unitemporal geometric dynamics*

$$m \frac{d^2 x}{dt^2} = e \frac{dx}{dt} \times B + \frac{e^2}{m} \nabla f_A - e \partial_t A, \quad B = \text{rot } A,$$

which is a movement in a gyroscopic field of forces [63]–[68], [69, 71, 72] or a B-vortex dynamics. The associated Hamiltonian is

$$H_2 = \frac{1}{2} \delta_{ij} \left(m \frac{dx^i}{dt} - eA^i \right) \left(m \frac{dx^j}{dt} + e A^j \right)$$

$$= \frac{m^2}{2} \delta_{ij} \frac{dx^i}{dt} \frac{dx^j}{dt} - e^2 f_A.$$

Remark 5.2: In general, unitemporal geometric dynamics produced by the potential vector "$-A$" is different from the classical Lorentz law because

$$L_2 - mL_1 = \frac{1}{2} e^2 \delta_{ij} A^i A^j + meV$$

and the force $\frac{e}{m} \nabla f_A - \partial_t A$ is not the electric field $E = -\nabla V - \partial_t A$. In other words, the Lagrangians L_1 and L_2 are not in the same class of Lagrangian equivalence.

5.5.2 Unitemporal Geometric Dynamics Produced by Magnetic Induction *B*

Because we want to analyze geometric dynamics using units of measure, magnetic flux is described by

$$m \frac{dx}{dt} = \lambda B,$$

where the unit of measure for the constant λ is $[kgm^3/Vs^2]$.

The least squares Lagrangian is

$$L_3 = \frac{1}{2} \delta_{ij} \left(m \frac{dx^i}{dt} - \lambda B^i \right) \left(m \frac{dx^j}{dt} - \lambda B^j \right)$$

$$= \frac{1}{2} \left\| m \frac{dx}{dt} - \lambda B \right\|^2.$$

It gives the Euler–Lagrange equations (*unitemporal magnetic geometric dynamics*)

$$m \frac{d^2x}{dt^2} = \lambda \frac{dx}{dt} \times \text{rot } B + \frac{\lambda^2}{m} \nabla f_B + \lambda \partial_t B,$$

where

$$f_B = \frac{1}{2} \delta_{ij} B^i B^j = \frac{1}{2} ||B||^2$$

is the *magnetic energy density*. This is actually a dynamic imposed by a gyroscopic force or a *J*-vortex.

The associated Hamiltonian is

$$H_3 = \frac{1}{2} \delta_{ij} \left(m \frac{dx^i}{dt} - \lambda B^i \right) \left(m \frac{dx^j}{dt} + \lambda B^j \right)$$

$$= \frac{m^2}{2} \delta_{ij} \frac{dx^i}{dt} \frac{dx^j}{dt} - \lambda^2 f_B.$$

5.5.3 Unitemporal Geometric Dynamics Produced by Electric Field *E*

Electric flow is described by

$$m \frac{dx}{dt} = \lambda E,$$

where the unit of measure for the constant λ is $[kgm^2/Vs]$. The least squares Lagrangian appears

$$L_4 = \frac{1}{2} \delta_{ij} \left(m \frac{dx^i}{dt} - \lambda E^i \right) \left(m \frac{dx^j}{dt} - \lambda E^j \right)$$

$$= \frac{1}{2} \left\| m \frac{dx}{dt} - \lambda E \right\|^2,$$

with Euler–Lagrange ODEs (*unitemporal electric geometric dynamics*)

$$m \frac{d^2 x}{dt^2} = \lambda \frac{dx}{dt} \times \operatorname{rot} E + \frac{\lambda^2}{m} \nabla f_E + \lambda \partial_t E,$$

where

$$f_E = \frac{1}{2} \delta_{ij} E^i E^j = \frac{1}{2} ||E||^2$$

is the *electric energy density*; here we actually have a dynamic in $\partial_t B$ -vortex. The associated Hamiltonian is

$$H_4 = \frac{1}{2} \delta_{ij} \left(m \frac{dx^i}{dt} - \lambda E^i \right) \left(m \frac{dx^j}{dt} + \lambda E^j \right)$$

$$= \frac{m^2}{2} \delta_{ij} \frac{dx^i}{dt} \frac{dx^j}{dt} - \lambda^2 f_E.$$

Problem 5.5.1: (Open problem) As is well known, charged particles in the earth's magnetic field spiral from pole to pole. Similar movements are observed in plasma laboratories and refer to electrons in a metal subjected to an external magnetic field. Until our work, these movements were justified only by the classical universal Lorentz law.

Can we justify these movements via geometric dynamics produced by appropriate vector fields, within the meaning of this paragraph or of papers [63],[68], [69, 71, 72]?

5.5.4 Potentials Associated to Electromagnetic Forms

Let $U \subset \mathbb{R}^3 = M$ be an isotropic, homogeneous and linear medium domain. What mathematical objects describe electromagnetic fields: vector fields or differential forms? It has been shown [3] that the magnetic field H and the electric field E are 1-forms; the magnetic induction B, electrical displacement D and the current density J are all 2-forms; the electric charge density ρ is a 3-form.

The operator d is the exterior derivative and the operator ∂_t is the partial derivative with respect to the time t.

In terms of differentiable forms, the *Maxwell PDEs* on $U \times \mathbb{R}$ can be written in the form

$$dD = \rho, \qquad dH = J + \partial_t D$$

$$dB = 0, \qquad dE = -\partial_t B.$$

The constitutive relations are

$$D = \varepsilon * E, \; B = \mu * H,$$

where the star operator $*$ means the Hodge operator, ε is the permittivity and μ is the scalar permeability.

The local components $E_i, i = 1, 2, 3$, of the 1-form E are called *electric potentials*, and the local components $H_i, i = 1, 2, 3$, of the 1-form H are called *magnetic potentials*.

5.5.5 Potential Associated to Electric 1-form E

Let us consider the function $V : \mathbb{R} \times U \to \mathbb{R}$, $(t, x) \mapsto V(t, x)$ and the Pfaff equation $dV = -E$ or equivalently the PDEs system $\dfrac{\partial V}{\partial x^i} = -E_i$, $i = 1, 2, 3$. Of course, the conditions of complete integrability $\dfrac{\partial^2 V}{\partial x^i \partial x^j} = \dfrac{\partial^2 V}{\partial x^j \partial x^i}$ require $dE = 0$ (*electrostatic field*), which is not always satisfied. In any case, we can introduce the least squares Lagrangian

$$L_5 = \frac{1}{2} \delta^{ij} \left(\frac{\partial V}{\partial x^i} + E_i \right) \left(\frac{\partial V}{\partial x^j} + E_j \right) = \frac{1}{2} ||dV + E||^2.$$

This Lagrangian produces the Euler–Lagrange PDE (Poisson equation)

$$\Delta V = -\operatorname{div} E,$$

and consequently V must be an *electric potential*. For a linear isotropic material, we have $D = \varepsilon E$, with $\rho = \operatorname{div} D = \varepsilon \operatorname{div} E$. We find the PDE of the potential for a homogeneous material ($\varepsilon = \text{constant}$)

$$\Delta V = -\frac{\rho}{\varepsilon}.$$

For the space without charges, we have $\rho = 0$, and then the potential V satisfies the Laplace equation $\Delta V = 0$ (*harmonic function*). The Lagrangian L_5 produces the Hamiltonian

$$H_5 = \frac{1}{2} \langle dV - E, dV + E \rangle,$$

and the momentum-energy tensor field

$$T^i_j = \frac{\partial V}{\partial x^j} \frac{\partial L}{\partial \left(\dfrac{\partial V}{\partial x^i} \right)} - L_5 \delta^i_j = \frac{\partial V}{\partial x^j} \left(\frac{\partial V}{\partial x^i} + E_i \right) - L_5 \delta^i_j.$$

5.5.6 Potential Associated to Magnetic 1-form H

Now, let us consider the Pfaff equation $d\varphi = H$ or the PDEs system $\dfrac{\partial \varphi}{\partial x^i} = H_i$, $i = 1, 2, 3$, $\varphi : \mathbb{R} \times U \to \mathbb{R}$, $(t, x) \mapsto \varphi(t, x)$. The conditions of complete integrability $dH = 0$ are satisfied only in particular cases. Building the least squares Lagrangian

$$L_6 = \frac{1}{2} \delta^{ij} \left(\frac{\partial \varphi}{\partial x^i} - H_i \right) \left(\frac{\partial \varphi}{\partial x^j} - H_j \right) = \frac{1}{2} ||d\varphi - H||^2,$$

we obtain the Euler–Lagrange PDE (Laplace equation) and hence φ must be the *magnetic potential*. The Lagrangian L_6 produces the Hamiltonian $H_6 = \dfrac{1}{2} \langle d\varphi - H, d\varphi + H \rangle$.

5.5.7 Potential Associated to Potential 1-form A

Since $dB = 0$, there exists a potential 1-form A satisfying $B = dA$. Now let us consider the Pfaff equation $d\psi = A$ or equivalently the PDEs system $\dfrac{\partial \psi}{\partial x^i} = A_i$, $i = 1, 2, 3$, ψ:

$\mathbb{R} \times U \to \mathbb{R}, (t, x) \mapsto \psi(t, x)$. The complete integrability condition $dA = 0$ is satisfied only if $B = 0$. Building the least squares Lagrangian

$$L_7 = \frac{1}{2} \delta^{ij} \left(\frac{\partial \psi}{\partial x^i} - A_i \right) \left(\frac{\partial \psi}{\partial x^j} - A_j \right)$$

$$= \frac{1}{2} ||d\psi - A||^2,$$

we obtain the Euler–Lagrange PDE (Laplace PDE) $\Delta \psi = 0$. The Lagrangian L_7 gives the Hamiltonian $H_7 = \frac{1}{2} \langle d\psi - A, d\psi + A \rangle$ and the momentum-energy tensor field

$$T^i{}_j = \frac{\partial \psi}{\partial x^j} \left(\frac{\partial \psi}{\partial x^i} - A_i \right) - L_7 \delta^i_j.$$

Problem 5.5.2: **(Open problem)** Find the physical interpretation of the extremals of least squares Lagrangians

$$L_8 = \frac{1}{2} ||dA - B||^2$$

$$= \frac{1}{2} \delta^{ik} \delta^{jl} \left(\frac{\partial A_i}{\partial x^j} - \frac{\partial A_j}{\partial x^i} - B_{ij} \right) \left(\frac{\partial A_k}{\partial x^l} - \frac{\partial A_l}{\partial x^k} - B_{kl} \right),$$

$$L_9 = \frac{1}{2} ||dE + \partial_t B||^2 + \frac{1}{2} ||dH - J - \partial_t D||^2$$

$$+ \frac{1}{2} ||dD - \rho||^2 + \frac{1}{2} ||dB||^2,$$

which are solutions of Maxwell PDEs.

Remark 5.3: There are many applications of previous theories. One of the most important is in strength of materials. In problems associated with the twisting of a cylinder or a prism, we have to investigate the double integral functional

$$J(z(\cdot)) = \int_D \left(\left(\frac{\partial z}{\partial x} - y \right)^2 + \left(\frac{\partial z}{\partial y} + x \right)^2 \right) dx dy$$

in the sense of finding an extremum. The Euler–Lagrange PDE $\frac{\partial^2 z}{\partial x^2} + \frac{\partial^2 z}{\partial y^2} = 0$ shows that the extremals are harmonic functions. Of course, J has no global minimum point because the PDEs system $\frac{\partial z}{\partial x} = y$, $\frac{\partial z}{\partial y} = -x$ is not completely integrable.

5.6 Wind Theory and Geometric Dynamics

A *wind* refers to the dynamic behavior of air under atmospheric disturbance conditions. Mathematically, the wind is represented by a PDE where the velocity vector field $\vec{v}(x, y, z, t)$, $(x, y, z) \in \mathbb{R}^3$, $t \in \mathbb{R}$, is unknown (see [11]). The disturbing forces that apply to air particles are: (i) *inertial force* (represented by the total acceleration) $\frac{D\vec{v}}{Dt} = \frac{\partial \vec{v}}{\partial t} + (\vec{v} \nabla) \vec{v}$;

(ii) *force derived from the horizontal pressure gradient* $-\frac{1}{\rho}$ grad p; (iii) *deviation force due to rotation of the earth* $\vec{v} \times \vec{F}_1$; (iv) *smoothing pressure* produced by the relative motion of air layers, up and down, $\vec{F}_2$.

The wind PDE (air movement, similar to Stokes Navier PDE for moving a fluid) is

$$\underbrace{\frac{D\vec{v}}{Dt}}_{(1)} = \underbrace{-\frac{1}{\rho} \, \text{grad} \, p}_{(2)} + \underbrace{\vec{v} \times \vec{F}_1}_{(3)} + \underbrace{\vec{F}_2}_{(4)} \, .$$

The winds are classified according to the relative importance of the four forces:

(1) *Geostrophic wind.* The movement is stationary (independent of time t). Only forces (2) and (3) remain:

$$-\frac{1}{\rho} \, \text{grad} \, p + \vec{v} \times \vec{F}_1 = 0.$$

This wind is due to the horizontal pressure gradient, the trajectories of the motion being *isobars*.

In atmospheric science, a geostrophic wind results from an exact balance between the Coriolis force and the pressure gradient force. This condition is called geostrophic equilibrium or geostrophic balance (also known as geostrophy). The geostrophic wind is directed parallel to isobars (lines of constant pressure at a given height).

(2) *Gradient wind.* The first force is approximated by $\frac{v^2}{R}$, where $v^2 = \langle \vec{v}, \vec{v} \rangle$, and R is curvature radius of the isobars. The partial acceleration $\frac{\partial \vec{v}}{\partial t}$ and the fourth force are omitted:

$$\frac{v^2}{R} = -\frac{1}{\rho} \, \text{grad} \, p + \vec{v} \times \vec{F}_1.$$

It remains that the pressure gradient is a quadratic function of the velocity.

The gradient wind is defined as a horizontal wind having the same direction as the geostrophic wind but with a magnitude consistent with a balance of three forces: the pressure gradient force, the Coriolis force and the centrifugal force arising from the curvature of a parcel trajectory.

(3) *Antitryptic wind.* Forces (2) and (4) are dominant. The force (4) represents the friction at ground level and thus is a force opposite to the direction of movement. The wind is evolving towards low pressure areas.

A wind for which the pressure force exactly balances the viscous force, in which the vertical transfers of momentum predominate.

(4) *Ageostrophic wind.* It moves away from the geostrophic wind by a frictionless movement. The term (4) is dominant and expresses convection or crawling at ground level.

Ageostrophic winds are merely the component of the actual wind that is not geostrophic. In other words, given the actual wind v and the geostrophic wind v_g, the ageostrophic wind v_a is the vector difference between them, i.e., $v_a = v - v_g$. The ageostrophic wind represents friction and other effects.

Let us replace the velocity vector field $\vec{v}(x, y, z, t)$ defined on $\mathbb{R}^3 \times \mathbb{R}$ by the velocity vector field $\vec{v}(t)$ along a curve $x = x(t), y = y(t), z = z(t)$ (defined on $\mathbb{R}$). Then the term $(\vec{v}\nabla)\vec{v}$ from the total acceleration does not exist. To simplify, we eliminate the arrow.

Definition 5.1: A vector solution $v(t)$ of the first-order ODE system of type

$$\dot{v}(t) = \operatorname{grad} f(x(t)) + \operatorname{grad} \varphi(x(t)) \times v(t)$$

is called wind along the curve $x(t)$.

The Geometric Dynamics on the Riemannian manifold $(\mathbb{R}^3, \delta_{ij})$ produces a special wind, using as unknown the velocity vector field $v(t)$ along a curve (defined on $\mathbb{R}$).

In the following, we will show that certain flows and Riemannian metrics generate winds via geometric dynamics technique. We exemplify by the pendulum flow and Lorenz flow (see also [63]–[68], [69, 71, 72]).

5.6.1 Pendular Geometric Dynamics and Pendular Wind

The small oscillations of a plane pendulum are the solutions of the differential system

$$\frac{dx^1}{dt} = -x^2, \quad \frac{dx^2}{dt} = x^1.$$

We need the Riemannian manifold $(\mathbb{R}^2, \delta_{ij})$ to create a least squares Lagrangian from this differential system and the Riemannian metric δ_{ij}.

In this case $x^1(t) = 0$, $x^2(t) = 0$, $t \in \mathbb{R}$, is the equilibrium point and $x^1(t) = c_1 \cos t + c_2 \sin t$, $x^2(t) = c_1 \sin t - c_2 \cos t$, $t \in \mathbb{R}$, is the general solution (family of concentric circles).

Let

$$X = (X^1, X^2), \ X^1(x^1, x^2) = -x^2, \ X^2(x^1, x^2) = x^1,$$

and its extension $(-x^2, x^1, 0)$ to $\mathbb{R}^3$. We find

$$f(x^1, x^2) = \frac{1}{2}((x^1)^2 + (x^2)^2), \ \operatorname{rot} X = (0, 0, 2), \ \operatorname{div} X = 0.$$

The pendulum flux preserves the area.

The prolongation by derivation is $\ddot{x}^i = \frac{\partial X^i}{\partial x^j} \dot{x}^j$ or $\ddot{x}^1 = -\dot{x}^2$, $\ddot{x}^2 = \dot{x}^1$. This prolongation has the general solution (family of circles)

$$x^1(t) = a_1 \cos t + a_2 \sin t + h, \ x^2(t) = a_1 \sin t - a_2 \cos t + k, \ t \in \mathbb{R}.$$

The pendular geometric dynamics is described by the differential system

$$\ddot{x}^i = \left(\frac{\partial X^i}{\partial x^j} - \frac{\partial X^j}{\partial x^i} \right) \dot{x}^j + \frac{\partial f}{\partial x^i}$$

or explicit

$$\ddot{x}^1 = x^1 - 2\dot{x}^2, \ \ddot{x}^2 = x^2 + 2\dot{x}^1,$$

with the general solution (family of spirals)

$$x^1(t) = b_1 \cos t + b_2 \sin t + b_3 t \cos t + b_4 t \sin t,$$

$$x^2(t) = b_1 \sin t - b_2 \cos t + b_3 t \sin t - b_4 t \cos t, \ t \in \mathbb{R}.$$

Let us now explain how a pendular (geostrophic) wind appears from a pendular current and a Euclidean metric. First, we extend the pendular vector field from $\mathbb{R}^2$ to $\mathbb{R}^3$, i.e., we set $X = (X^1, X^2, X^3)$ with $X^1(x^1, x^2, x^3) = -x^2$, $X^2(x^1, x^2, x^3) = x^1$, $X^3(x^1, x^2, x^3) = 0$. The vector field $\operatorname{rot} X = (0, 0, 2)$ can be written in the form $\operatorname{grad} \varphi$, where $\varphi(x^1, x^2, x^3) = 2x^3$. Consequently, the pendular wind is described by the differential system

$$\ddot{x} = \operatorname{grad} f + \operatorname{grad} \varphi \times \dot{x},$$

which transforms into

$$\dot{v} = \operatorname{grad} f + \operatorname{grad} \varphi \times v.$$

5.6.2 Lorenz Geometric Dynamics and Lorenz Wind

The Lorenz flow is the first dissipative model with chaotic behavior discovered through numerical experiments. The differential equations of state are

$$\dot{x}^1 = -\sigma x^1 + \sigma x^2, \ \dot{x}^2 = -x^1 x^3 + r x^1 - x^2, \ \dot{x}^3 = x^1 x^2 - b x^3,$$

where σ, r, b are real parameters.

We use the Riemannian manifold $(\mathbb{R}^3, \delta_{ij})$ to create a least squares Lagrangian from Lorenz flow and the Riemannian metric δ_{ij}.

Usually the parameters σ, b are kept fixed while r is considered variable. Chaotic behavior is observed for

$$r > r_0 = \frac{\sigma(\sigma + b + 3)}{\sigma - b - 1}.$$

With $\sigma = 10, b = \frac{8}{3}$, the previous equality gives $r_0 = 24,7368$. If $\sigma \neq 0$ and $b(r-1) > 0$, then the equilibrium points of the Lorenz flow are

$$x^1 = 0, \ x^2 = 0, \ x^3 = 0$$

$$x^1 = \pm\sqrt{b(r-1)}, \ x^2 = \pm\sqrt{b(r-1)}, \ x^3 = r - 1.$$

Let $X = (X^1, X^2, X^3)$ with

$$X^1 = -\sigma x^1 + \sigma x^2, \ X^2 = -x^1 x^3 + r x^1 - x^2, \ X^3 = x^1 x^2 - b x^3$$

and

$$2f = (-\sigma x^1 + \sigma x^2)^2 + (-x^1 x^3 + r x^1 - x^2)^2 + (x^1 x^2 - b x^3)^2$$

$$\operatorname{rot} X = (2x^1, -x^2, r - x^3 - \sigma).$$

The geometric dynamics is described by the second-order system

$$\ddot{x}^1 = \frac{\partial f}{\partial x^1} + (\sigma + x^3 - r)\dot{x}^2 - x^2\dot{x}^3$$

$$\ddot{x}^2 = \frac{\partial f}{\partial x^2} + (-\sigma - x^3 + r)\dot{x}^1 - 2x^1\dot{x}^3$$

$$\ddot{x}^3 = \frac{\partial f}{\partial x^3} + x^2\dot{x}^1 + 2x^1\dot{x}^2.$$

Let us now explain how a Lorenz geostrophic wind appears (produced by the Lorenz flow and the Euclidean metric). Since $\Delta X = 0$, i.e. $\Delta X^i = 0$, the result is $\Delta f \geq 0$, i.e., f is a subharmonic function. Therefore the critical points of f cannot be maximum points.

We remark that $\mathrm{rot}\, X$ can be written in the form $\mathrm{grad}\,\varphi$, where

$$\varphi = (x^1)^2 - \frac{1}{2}(x^2)^2 - \frac{1}{2}(x^3 + \sigma - r)^2.$$

Therefore the Lorenz wind is described by the differential system

$$\ddot{x} = \mathrm{grad}\, f + \mathrm{grad}\,\varphi \times \dot{x}$$

or

$$\dot{v} = \mathrm{grad}\, f + \mathrm{grad}\,\varphi \times v.$$

5.7 Maple Application Topics

Problem 5.7.1: Find the primitives of the coefficients of Van der Pol equation (oscillators)

$$\ddot{x}(t) - \mu(1 - x(t)^2)\dot{x}(t) + x(t) = 0.$$

Using the Maple command "DEtools[DEplot] - plot solutions to a system of DEs", > with(DEtools): > DEplot($de, x(t), t = 0..15, x = -1..1, [[x(0) = 1, (D(x))(0) = 0]]$) find the image of a solution.

Solution. The equality $f'(x) = x$ implies $f(x) = \frac{x^2}{2}$. The condition $\varphi'(x) = -\mu(1 - x^2)$ gives $\varphi(x) = -\mu\left(x - \frac{x^3}{3}\right)$.

Problem 5.7.2: Find the primitives of the coefficients of Duffing equation (oscillators)

$$\ddot{x}(t) + \mu\dot{x}(t) + \alpha x(t) + \beta x(t)^3 = \gamma\cos(\omega t).$$

Using the Maple command "DEtools[DEplot] - plot solutions to a system of DEs", > with(DEtools): > DEplot($de, x(t), t = 0..15, x = -1..1, [[x(0) = 1, (D(x))(0) = 0]]$) find the image of a solution.

Problem 5.7.3: Write the coefficients of Poisson–Boltzmann equation (Statistical Physics)

$$\ddot{x}(t) + \frac{\alpha}{t}\,\dot{x}(t) = e^{x(t)}$$

as derivatives. Using the Maple command "DEtools[DEplot] - plot solutions to a system of DEs", $>$ with(DEtools): $>$ DEplot$(de, x(t), t = 0..15, x = -1..1, [[x(0) = 1, (D(x))(0) = 0]])$ find the image of a solution.

Problem 5.7.4: **(Spring pendulum)** Consider a pendulum made of a spring with a mass m on the end. The spring is arranged to lie in a straight line (which we can arrange by wrapping the spring around a rigid massless rod). The equilibrium length of the spring is ℓ. Let the spring have the length $\ell + x(t)$, and let its angle with the vertical be $\theta(t)$. Assuming that the motion takes place in a vertical plane, find the equations of motion for x and θ.

Solution. The kinetic energy may be broken up into the radial and tangential parts,

$$T = \frac{1}{2}m(\dot{x}^2 + (\ell + x)^2\dot{\theta}^2).$$

The potential energy comes from both gravity and the spring,

$$V = -g(\ell + x)\cos\theta + \frac{1}{2}kx^2.$$

It follows the Lagrangian $L = T - V$ and the Euler–Lagrange ODEs.

Problem 5.7.5: Let $A = |A|e^{i\phi}$ be a complex quantity describing the disturbance, $\sigma = \sigma_1 + i\sigma_2$ be the complex growth rate, and $\ell = \ell_1 + i\ell_2$ be a complex number, where ℓ_1 is the Landau constant. The Stuart–Landau ODE

$$\frac{dA}{dt} = \sigma A - \frac{\ell}{2}A|A|^2$$

describes the behavior of a nonlinear oscillating system near the Hopf bifurcation, named after John Trevor Stuart and Lev Landau.

Write the associated least squares Lagrangian and the Stuart–Landau wind.

Problem 5.7.6: Find the extremals of the double integral functional

$$J(z(\cdot)) = \int_D \left(\left(\frac{\partial z}{\partial x} - y\right)^2 + \left(\frac{\partial z}{\partial y} + x\right)^2 \right) dxdy$$

constrained by PDE

$$x\frac{\partial z}{\partial x} - y\frac{\partial z}{\partial y} = 2axy(x^2 - y^2).$$

Problem 5.7.7: Find the extremals of the double integral functional

$$J(z(\cdot)) = \int_D \left(\left(\frac{\partial z}{\partial x} - y \right)^2 + \left(\frac{\partial z}{\partial y} + x \right)^2 \right) dx dy$$

constrained by PDE

$$\left(\frac{\partial z}{\partial x} \right)^3 + \left(\frac{\partial z}{\partial y} \right)^3 = 3a \frac{\partial z}{\partial x} \frac{\partial z}{\partial y}.$$

Problem 5.7.8: **(Poincaré–Cartan invariant)** Fix $t \in \mathbb{R}$ and consider a closed curve $s \mapsto \gamma(s) = (x(s,t), p(s,t)) : [0,T] \to \mathbb{R}^{2n}$. We add the Hamiltonian $H(x,p)$. Suppose that for each fixed $s \in [0,1]$, we have

$$\frac{d}{dt}x(s,t) = -D_p H(x(s,t), p(s,t)), \quad \frac{d}{dt}p(s,t) = D_x H(x(s,t), p(s,t)).$$

Show that the curvilinear integral $\oint p_i \, dx^i = \int_0^1 p_i \frac{\partial x^i}{\partial s} \, ds$ is independent of t.

Problem 5.7.9: **(Maupertuis principle)** Consider a system with an autonomous Lagrangian L and energy (Hamiltonian) $H(x, \dot{x}) = \langle \dot{x}, D_{\dot{x}}L(x, \dot{x}) \rangle - L(x, \dot{x})$. The energy is conserved by the solutions of the Euler–Lagrange ODEs. Show the critical points of the action are also critical points of the functional $\int_0^T (L + H) \, dt = \int_0^T \langle \dot{x}, D_{\dot{x}}L(x, \dot{x}) \rangle \, dt$ under the constraint that energy is conserved.

6

Nonholonomic Constraints

Motto:
"Flocks of darkling starlings climb the sky Westwards,
Like the sickly plumage of the gray hornbeam
Which is moulting sadly, as it casts upwards
Leaves in azure's beam."
Tudor Arghezi – *Never Yet Had Autumn...*

In this chapter, we present some selected applications from science and engineering: rolling cylinders as models with holonomic constraints; rolling disc (unicycle), four-wheeled robot, the trailer with several components, as models with nonholonomic constraints. The nonholonomic models are described in spaces with at least three dimensions, the restrictions being nonholonomic manifolds in the sense of Vranceanu or geometric distributions. Traditional texts stop only at extremum necessary conditions, since the treated problems certainly have solutions and once found, they are and not others. For details, see [2, 5, 23, 38, 39, 50, 72, 73, 78].

6.1 Models With Holonomic and Nonholonomic Constraints

We start with the classification of dynamical systems described by ODEs or PDEs.

Definition 6.1: Suppose a dynamical system is described by ODEs of the type $F^a(x(t), \dot{x}(t), t) = 0$, $x = (x^1, ..., x^n) \in \mathbb{R}^n$, $t \in \mathbb{R}$, $a = \overline{1, m}$, $m \leq n$, the solutions being curves $x(t)$. If this system of ODEs can be replaced by the algebraic system $G^a(x, t) = c^a$, $x = (x^1, ..., x^n) \in \mathbb{R}^n$, $t \in \mathbb{R}$, $a = \overline{1, m}$, $m \leq n$, with the property $G^a(x(t), t) \equiv c^a$, $dG^a(x(t), t) = 0$, $\forall t \in \mathbb{R}$, then the functions $G^a(x, t)$ are called *first integrals*, the constraints $G^a(x, t) = c^a$ are called *holonomic* and the dynamical system is called *holonomic*. Otherwise, the constraints in velocities $F^a(x(t), \dot{x}(t), t) = 0$ are called *nonholonomic*, and the dynamical system is called *nonholonomic*.

A similar definition can be formulated for PDEs.

A nonholonomic system in physics and mathematics is a physical system whose state depends on the path taken in order to achieve it.

Definition 6.2: A dynamical system on $\mathbb{R}^n$ is called: (i) scleronom (i.e. rigid), or autonomous, if the equations not explicitly dependent on time t; (ii) reonom (i.e. flowing), or non-autonomous, if the equations explicitly depend on time t.

In many cases the constraints are given by *Pfaff equations*

$$\lambda_i^a(x)dx^i = 0, \ a = \overline{1,m}, \ i = \overline{1,n}, \ m < n.$$

If $m = n - 1$, then the constraints are holonomic (for example, the case $n = 2$, $m = 1$ is always holonomic). If $m < n - 1$, then the constraints can be holonomic or nonholonomic. The integral manifolds of previous Pfaff system can be at most $(n - m)$-dimensional. If we have only $(n - m)$-dimensional solutions, then the system is holonomic.

In nonholonomic Pfaff case, the maximum dimension of integral manifold is not obvious, but it cannot overcome $n - m - 1$. On the nonholonomic case, Cauchy problems attached to nonholonomic constraints (particularly, Pfaff equations) do not have unique solutions because the tangent space at a point to an integral manifold is strictly included in the tangent space to the constraint (particularly, distribution generated by Pfaff equations).

A nonholonomic system of Pfaff type ($n \geq 3$) has sure one-dimensional solutions (curves); this is almost obvious because dividing by dt, the Pfaff equations turn into ordinary differential equations linear in velocities, with n unknown functions, $\lambda_i^a(x(t))\dot{x}^i(t) = 0$.

It is not necessary for all nonholonomic constraints to take the Pfaff form. Generally, they may involve higher derivatives or inequalities. In this context, we accept that the solutions of nonholonomic constraints are curves.

A geometric meaning of nonholonomic constraints is such that they represent submanifolds in the jet space J^1.

Example 6.1: *The motion of a mechanical system is for $t > 0$ subject to the following nonholonomic constraint $\dot{q}^1(t)^2 + \dot{q}^2(t)^2 + \dot{q}^3(t)^2 - \frac{1}{t} = 0$, meaning that the particle's speed decreases proportionally to $\frac{1}{\sqrt{t}}$. This nonholonomic constraint is rheonomic and is affine of degree 2 in components of velocity.*

Sources for movements with nonholonomic constraints can be classified as follows:

(i) *bodies in rolling contact without slipping:*
- mobile robots with wheels or cars (wheels that roll on the ground without skidding or slipping);
- dexterous handling with multi-fingered robot arms (fingertips around grasped objects);

(ii) *conservation of angular moments in multibody systems:*
- robotic manipulators floating in space (without any external action);
- dynamically balanced robots, trampoline jumpers or astronauts (in flight or gliding phases);
- satellites with reaction wheels (or torque) to stabilize the position;

(iii) *special control operations:*
  - non-cyclic inversion schemes for redundant robots (m load commands for n constraints);
  - robotic systems floating under water ($m = 4$ input speeds for $n = 6$ generalized coordinates).

Explanatory note The optimization of an integral functional subject to holonomic or nonholonomic constraints is called Hamilton's principle. There are cases of mechanical problems with nonholonomic constraints, in which the mathematical solution according to Hamilton's principle does not make mechanical sense.

Here's why: another principle is needed to derive the equations of motion for problems with nonholonomic constraints in mechanics. It is called d'Alembert's Principle, which states that total virtual work of the forces is zero for all (reversible) variations that satisfy the given kinematical conditions. We can think d'Alembert's Principle as the condition $\delta L = 0$.

For holonomic problems one has the equality

$$\int_{t_0}^{t_1} \delta L \, dt = \delta \int_{t_0}^{t_1} L \, dt,$$

while for some nonholonomic problems the operators "integral and δ" do not commute, i.e.,

$$\int_{t_0}^{t_1} \delta L \, dt \neq \delta \int_{t_0}^{t_1} L \, dt.$$

(see [10]; [51], §26.1; [75], pp. 132-133; [83], §87).

Some nonholonomic constraints in mechanics are, typically, of the Pfaff form

$$g_k(t, \mathbf{q}, \dot{\mathbf{q}}) = \sum_{j=1}^{n} a_{kj}(t, \mathbf{q}) \dot{q}_j = 0, \quad k = \overline{1, m}.$$

For the nonholonomic problems with m constraints as above, d'Alembert's Principle yields equations of the form

$$\frac{d}{dt} \frac{\partial L}{\partial \dot{q}_j} - \frac{\partial L}{\partial q_j} = \sum_{k=1}^{m} \mu_k(t) a_{kj}(t, \mathbf{q}),$$

where L is (unmodified) Lagrangian, i.e., $L = T - V$, and the functions μ_k are multipliers to be determined using the n Euler–Lagrange equations and the m differential equations.

Let us start from the theory of extrema with nonholonomic constraints of Pfaff type, developed by us in [72, 73]. Extending to the problem "extrema of the functional $\int_{t_0}^{t_1} L(t, \mathbf{q}(t), \dot{\mathbf{q}}(t)) dt$ subject to nonholonomic Pfaff equations $a_{kj}(t, \mathbf{q}) \dot{q}^j = 0$", we obtain the d'Alembert's principle $\mathcal{EL}_j = v^k(t) a_{kj}(t, \mathbf{q})$ (see also [45]).

In general, the d'Alembert ODEs system is not equivalent to the Euler–Lagrange ODEs system associated to the Lagrangian $L - (\lambda_1(t) g_1 + \cdots + \lambda_m(t) g_m)$, resulting from Hamilton's Principle.

Example 6.2:   *Suppose a particle with mass m = 1 starting from origin* $(0, 0, 0)$ *with initial velocity* $(u, 0, w)$ *and moving constrained under single (nonholonomic) condition*

$$z\,dx - dy = 0.$$

Introducing $L = \frac{1}{2}(\dot{x}^2 + \dot{y}^2 + \dot{z}^2)$ *and applying d'Alembert's Principle, we obtain the equations*

$$\ddot{x}(t) = \lambda(t)z(t), \quad \ddot{y}(t) = -\lambda(t), \quad \ddot{z}(t) = 0,$$

in addition with equation of constraint $z(t)\dot{x}(t) - \dot{y}(t) = 0$. *The integration of this ODEs system gives the solution*

$$x(t) = \frac{u}{w}\ln\left(wt + \sqrt{1 + w^2t^2}\right), \; y(t) = \frac{u}{w}\left(\sqrt{1 + w^2t^2} - 1\right), \; z(t) = wt,$$

with $w \neq 0$ *and* $\lambda(t) = -\frac{1}{2}\dfrac{uw}{(1 + w^2t^2)^{3/2}}$. *Noting* $\sinh\theta = wt$, *the solution becomes*

$$x = \frac{u}{w}\theta, \quad y = \frac{u}{w}(\cosh\theta - 1), \quad z = \sinh\theta.$$

The particle will evolve on the surface parameterized by $\dfrac{u}{w}$ *and* θ.

But the previous motion is not a solution for the problem: extremize $\displaystyle\int_{t_0}^{t_1} L\,dt$ *subject to* $z\dot{x} - \dot{y} = 0$. *Indeed, Lagrange's multiplier theorem in Hamilton principle requires*

$$\mathcal{L} = \frac{1}{2}(\dot{x}^2 + \dot{y}^2 + \dot{z}^2) - \mu(t)(z\dot{x} - \dot{y}),$$

and the Euler–Lagrange ODEs are

$$\frac{d}{dt}(\dot{x} - \mu z) = 0, \quad \frac{d}{dt}(\dot{y} + \mu) = 0, \quad \ddot{z} + \mu\dot{x} = 0 \quad and \quad z\dot{x} - \dot{y} = 0.$$

The solution of this differential system can be found in Problem 7.6.1. The d'Alembert's motion does not satisfy these equations for any choice of μ.

These considerations can be summarized as follows:

Holonomic kinematical conditions can be attacked in *two ways.* If there are m equations between n variables, we can eliminate m of these and reduce the problem to $n - m$ independent variables. Or we can operate with $n + m$ variables and the given relations as auxiliary conditions.

Nonholonomic conditions require the *second* way, only. A reduction of variables is here not possible and we have to operate with more variables than the degrees of freedom of the system demand. The configurations space is here imbedded in a higher-dimensional space but without forming a definite subspace of it, because the kinematical conditions prescribe certain pencils of directions, but these directions do not envelop any surface.

From the viewpoint of the variational principles of mechanics, holonomic and nonholonomic conditions display a different behavior. Although the equations of motion can be extended to the case of nonholonomic conditions, yet these equations are not to be derivable from the principle that the variation of certain quantity vanishes.

6.2 Rolling Cylinder as a Model with Holonomic Constraints

Let us consider a fixed cylinder, of radius r_1. A second cylinder of radius $r_2 < r_1$ and mass m rolls without slipping over the first, the contact being along a generator. At a moment the outer cylinder peels off the first.

Problem 6.2.1: (Rolling cylinder, RC) Find the equations of motion of the outer cylinder. Specify the angle θ_1^* at which the cylinders separate (see Fig. 6.1).

Solution. We use the following coordinates: the angle θ_1 measured in clockwise and the angle θ_2 measured trigonometrically from the vertical through the centers of the cylinders; distance r from the center of the fixed cylinder to the center of the movable cylinder. The tangent point P moves to the point Q, when θ_1 changes. Because the small cylinder rolls without slipping over the largest, a first constraint is "the same arc length", i.e.,

$$s = \text{arc length} = r_1\theta_1 = r_2(\theta_2 - \theta_1).$$

The second constraint expresses the fact that during the rolling the two cylinders must be in contact, that is $r = r_1 + r_2 = \text{constant}$.

In short, the holonomic constraints are

$$(r_1 + r_2)\theta_1 - r_2\theta_2 = 0, \ r - (r_1 + r_2) = 0.$$

The kinetic energy of the small cylinder consists of two parts: (1) the translational component containing the change $s = r\theta_1$; (2) the rotational component $\frac{1}{2}I\omega^2$ which contains the inertial moment $I = \frac{1}{2}mr_2^2$ of the small cylinder and the angular velocity $\omega = \dot\theta_2$. In this way the kinetic energy is

Figure 6.1 Rolling cylinders.

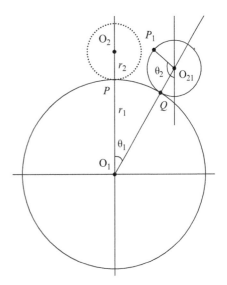

$$T = \frac{1}{2} m(r^2 \dot{\theta}_1^2) + \frac{1}{2} \left(\frac{1}{2} mr_2^2 \right) \dot{\theta}_2^2 \, .$$

$$\underbrace{\phantom{\frac{1}{2} m(r^2 \dot{\theta}_1^2)}}_{\text{translational}} \quad \underbrace{\phantom{\frac{1}{2} \left(\frac{1}{2} mr_2^2 \right) \dot{\theta}_2^2}}_{\text{rotational}}$$

On the other hand, the potential energy writes

$$V = mgh + \text{constant} = mgr \cos \theta_1 + \text{constant}.$$

It follows the Lagrangian

$$L = T - V = \frac{1}{2} m(r^2 \dot{\theta}_1^2) + \frac{1}{2} \left(\frac{1}{2} mr_2^2 \right) \dot{\theta}_2^2 - mgr \cos \theta_1$$

and the Euler–Lagrange equations of movement

$$\ddot{\theta}_1 + \frac{g}{r} \sin \theta_1 + \frac{\lambda}{mr} = 0, \quad \ddot{\theta}_2 + \frac{2\lambda}{mr}, \quad r\theta_1 = r_2 \theta_2,$$

which lead to $\left(1 - \dfrac{r}{2r_2} \right) \ddot{\theta}_1 + \dfrac{g}{r} \sin \theta_1 = 0.$

6.3 Rolling Disc (Unicycle) as a Model with Nonholonomic Constraint

A vertical disc (or wheel) that rolls on a plane (without slipping) follows a trajectory described in generalized coordinates (x, y, θ), where (x, y) is the contact point, and θ is the angle between Ox axis and direction of unicycle (see Fig. 6.2).

The nonholonomic constraint is $\tan \theta = \frac{\dot{y}}{\dot{x}}$, i.e., $\dot{x} \sin \theta - \dot{y} \cos \theta = 0$. Any configuration (x_f, y_f, θ_f) can be touched by the disc following the three steps: (1) rotate the disc until you aim the point (x_f, y_f); (2) roll the disc until we reach the point (x_f, y_f); (3) rotate the disc until it reaches the orientation θ_f.

The nonholonomic constraint (Pfaff equation)

$$dx \cdot \sin \theta - dy \cdot \cos \theta + d\theta \cdot 0 = 0$$

defines a field $\mathcal{H}$ of 2-planes

$$\mathcal{H}_{(x,y,\theta)} = \{(v_1, v_2, v_3) \mid v_1 \sin \theta - v_2 \cos \theta = 0\},$$

i.e., a two-dimensional non-integrable distribution, on the three-dimensional manifold $\mathbb{R}^2 \times S^1$.

6.3.1 Nonholonomic Geodesics

In unicycle case the generalized coordinates are (x, y, θ) (see Fig. 6.2). The nonholonomic restriction is the distribution $H : \dot{x} \sin \theta - \dot{y} \cos \theta = 0$. The normal vector field to the distribution is $N = (\sin \theta, -\cos \theta, 0)$. The unit vector fields $X_1 = (\cos \theta, \sin \theta, 0)$ and $X_2 = (0, 0, 1)$ are tangent to the distribution being orthogonal to N. The pair (X_1, X_2) is an orthonormal frame with respect to the Riemannian metric δ_{ij}.

Figure 6.2 Unicycle seen from above.

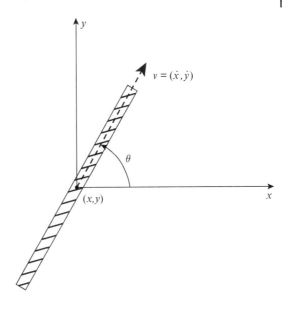

A trajectory $c(t) = (x(t), y(t), \theta(t))$ of contact point of the disc is tangent to the contact distribution since the tangent vector field $\dot{c} = \dot{x} \dfrac{\partial}{\partial x} + \dot{y} \dfrac{\partial}{\partial y} + \dot{\theta} \dfrac{\partial}{\partial \theta}$ can be re-written $(\dot{x} \cos \theta + \dot{y} \sin \theta) X_1 + \dot{\theta} X_2$. This conclusion is obtained from the identity $\lambda X_1 + \dot{\theta} X_2 \equiv (\dot{x}, \dot{y}, \dot{\theta})$.

The curve $c(t)$ is called *nonholonomic geodesic* if it is a solution to the following problem. Locally, such a trajectory is the shortest path between two given points.

Problem 6.3.1: (Variational problem of nonholonomic geodesics) Find

$$\min J(x(\cdot), y(\cdot), \theta(\cdot)) = \frac{1}{2} \int_{t_0}^{t_1} \left((\dot{x}(t) \cos \theta(t) + \dot{y}(t) \sin \theta(t))^2 + \dot{\theta}^2(t) \right) dt$$

with nonholonomic constraint

$$\dot{x}(t) \sin \theta(t) - \dot{y}(t) \cos \theta(t) = 0,$$

and boundary conditions

$$x(t_0) = x_0, \ \ y(t_0) = y_0, \ \ \theta(t_0) = \theta_0; x(t_1) = x_1, \ \ y(t_1) = y_1, \ \ \theta(t_1) = \theta_1.$$

Solution. To solve this extremum problem with nonholonomic constraint, we use the Lagrange multiplier rule. Entering the multiplier $p(t)$, we build a new Lagrangian

$$L = \frac{1}{2} \left((\dot{x} \cos \theta + \dot{y} \sin \theta)^2 + \dot{\theta}^2 \right) + p(t)(\dot{x} \sin \theta - \dot{y} \cos \theta),$$

changing the constraint optimization problem to a free one

$$\min \int_{t_0}^{t_1} L(\dot{x}(t), \dot{y}(t), \theta(t), \dot{\theta}(t), p(t)) \, dt.$$

If $(x^*(t), y^*(t), \theta^*(t), p^*(t))$ is a solution of the free optimization problem, then $(x^*(t), y^*(t), \theta^*(t), p^*(t))$ is a solution of the system

$$\frac{\partial L}{\partial x} - \frac{d}{dt}\frac{\partial L}{\partial \dot{x}} = 0, \quad \frac{\partial L}{\partial y} - \frac{d}{dt}\frac{\partial L}{\partial \dot{y}} = 0, \quad \frac{\partial L}{\partial \theta} - \frac{d}{dt}\frac{\partial L}{\partial \dot{\theta}} = 0,$$

$$\dot{x}(t)\sin\theta(t) - \dot{y}(t)\cos\theta(t) = 0,$$

made of Euler–Lagrange ODEs (including the nonholonomic constraint). Introducing the function $\eta = \dot{x}\cos\theta + \dot{y}\sin\theta$, we find

$$\frac{\partial L}{\partial x} = 0, \quad \frac{\partial L}{\partial \dot{x}} = \eta(t)\cos\theta(t) + p(t)\sin\theta(t),$$

$$\frac{\partial L}{\partial y} = 0, \quad \frac{\partial L}{\partial \dot{y}} = \eta(t)\sin\theta(t) - p(t)\cos\theta(t).$$

The first two Euler–Lagrange equations become

$$\eta(t)\cos\theta(t) + p(t)\sin\theta(t) = c_1, \quad \eta(t)\sin\theta(t) - p(t)\cos\theta(t) = c_2$$

or

$$\eta = c_1\cos\theta + c_2\sin\theta, \quad p = c_1\sin\theta - c_2\cos\theta.$$

Since $\frac{d\eta}{d\theta} = -p$, we find

$$\frac{\partial L}{\partial \theta} = \eta(t)\frac{d\eta}{d\theta}(t) + p(t)\eta(t) = 0, \quad \frac{\partial L}{\partial \dot{\theta}} = \dot{\theta}(t)$$

and thus the third Euler–Lagrange equation is reduced to $\ddot{\theta}(t) = 0$, with general solution $\theta(t) = at + b$.

Writing η in both ways and using the nonholonomic restriction, that is

$$\dot{x}\cos\theta + \dot{y}\sin\theta = c_1\cos\theta + c_2\sin\theta, \quad \dot{x}\sin\theta - \dot{y}\cos\theta = 0,$$

we deduce

$$\dot{x} = c_1\cos^2\theta + c_2\sin\theta\cos\theta, \quad \dot{y} = c_2\sin^2\theta + c_1\sin\theta\cos\theta, \quad \theta(t) = at + b.$$

Suppose $a \neq 0$. By integration we find the equations of geodesics in the form of a family of cycloids

$$x(t) = x_0 + \frac{c_1}{2a}(\sin(at+b)\cos(at+b) + at + b) + \frac{c_2}{2a}\sin^2(at+b),$$

$$y(t) = y_0 + \frac{c_2}{2a}(-\sin(at+b)\cos(at+b) + at + b) + \frac{c_1}{2a}\sin^2(at+b).$$

The constants are determined by imposing the conditions at the boundary. For particular values $(x_0 = y_0 = 0, a = \frac{1}{2}, b = 0, c_1 = 1, c_2 = 2)$, we find the cycloid in Fig. 6.3.

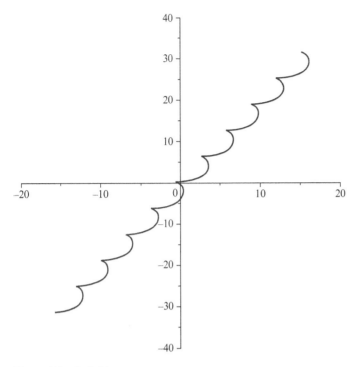

Figure 6.3 Cycloid.

6.3.2 Geodesics in Sleigh Problem

Let us refer to the *sleigh problem* or *skate problem*. We consider the movement of a simplified sled of unitary mass and unitary inertial momentum in $\mathbb{R}^2 \times S^1$, with coordinates (x, y, θ), subject to the nonholonomic constraint $\dot{y} = \dot{x} \tan \theta$. The Lagrangian of the problem is the kinetic energy

$$L = \frac{1}{2}(\dot{x}^2 + \dot{y}^2 + \dot{\theta}^2).$$

Its restriction to nonholonomic manifold $\dot{y} = \dot{x} \tan \theta$ is $\mathcal{L} = \frac{1}{2}(\dot{x}^2\sec^2\theta + \dot{\theta}^2)$. We can't escape the restriction, so we need a new Lagrangian

$$L_1 = \frac{1}{2}(\dot{x}^2 \sec^2 \theta + \dot{\theta}^2) + q(t)(\dot{y} - \dot{x}\tan\theta).$$

Using the Legendre transformation

$$\dot{x} = p_x \cos^2 \theta + p_y \sin\theta \cos\theta, \quad p_y = q(t), \quad \dot{\theta} = p_\theta,$$

we obtain the Hamiltonian

$$H_1 = \dot{x}\frac{\partial L_1}{\partial \dot{x}} + \dot{y}\frac{\partial L_1}{\partial \dot{y}} + \dot{\theta}\frac{\partial L_1}{\partial \dot{\theta}} - L_1,$$

i.e.,

$$H_1 = \frac{1}{2}(p_x \cos^2\theta + p_y \sin\theta \cos\theta)^2 \sec^2\theta - p_y \tan\theta + \dot{y}p_y + \frac{1}{2}p_\theta^2$$
$$+ (p_x \cos\theta + p_y \sin\theta)p_y \sin\theta$$

and the Hamilton ODEs

$$\dot{x} = \frac{\partial H_1}{\partial p_x}, \ \dot{y} = \frac{\partial H_1}{\partial p_y}, \ \dot{\theta} = \frac{\partial H_1}{\partial p_\theta}, \ \dot{p}_x = -\frac{\partial H_1}{\partial x}, \ \dot{p}_y = -\frac{\partial H_1}{\partial y}, \ \dot{p}_\theta = -\frac{\partial H_1}{\partial \theta},$$

to which we must add the restriction (Pfaff equation).

6.3.3 Unicycle Dynamics

We repeat (see Fig. 6.2): (i) in the case of unicycle the generalized coordinates are (x, y, θ); (ii) the nonholonomic constraint (non-integrable distribution) is

$$\dot{x} \sin\theta - \dot{y} \cos\theta = 0.$$

The vector field normal to the distribution is $N = (\sin\theta, -\cos\theta, 0)$. The vector fields $X_1 = (\cos\theta, \sin\theta, 0)$ and $X_2 = (0, 0, 1)$ are orthogonal to N, i.e., they are tangent to the distribution. The pair (X_1, X_2) is an orthonormal frame. Entering linear velocity (driving velocity) u^1 and angular velocity (steering velocity) u^2, it follows the kinematic model

$$(\dot{x}(t), \dot{y}(t), \dot{\theta}(t)) = u^1 X_1 + u^2 X_2.$$

Let m be the unicycle mass, I be the inertia momentum around vertical axis at contact point, u^1 driving force and u^2 steering torque. It follows the kinetic energy density (Lagrangian)

$$L = \frac{1}{2}\left(m\dot{x}^2(t) + m\dot{y}^2(t) + I\dot{\theta}^2(t)\right).$$

The general dynamic model (Euler–Lagrange ODEs system) is

$$(m\ddot{x}(t), m\ddot{y}(t), I\ddot{\theta}(t)) = \lambda N + \tau^1 X_1 + \tau^2 X_2,$$
$$\dot{x} \sin\theta - \dot{y} \cos\theta = 0.$$

This second-order differential system is changed in the first-order differential system

$$\dot{x} = (\cos\theta)\,v^1, \ \dot{y} = (\sin\theta)\,v^2, \ \dot{\theta} = v^2, \ m\dot{v}^1 = \tau_1, \ I\dot{v}^2 = \tau_2.$$

Setting $x = (x, y, \theta, v^1, v^2)$, this system can be written in the form of a first-order dynamical system

$$\dot{x} = f(x) + \tau^1 X_1 + \tau^2 X_2.$$

6.4 Nonholonomic Constraints to the Car as a Four-wheeled Robot

For simplification, we accept the "bicycle model": front and rear wheels collapse into two wheels at axle midpoints (see Fig. 6.4).

The generalized coordinates are (x, y, θ, ϕ), where ϕ = steering angle. It follows the noholonomic constraints (non-integrable distribution)

$$\dot{x}_f \sin(\theta + \phi) - \dot{y}_f \cos(\theta + \phi) = 0, \qquad \text{((front wheels))}$$

$$\dot{x} \sin\theta - \dot{y} \cos\theta = 0, \qquad \text{((rear wheels))}$$

on the four-dimensional manifold $\mathbb{R}^2 \times S^1 \times S^1$.

Being given the position of front wheels

$$x_f = x + \ell\cos\theta, \; y_f = y + \ell\sin\theta,$$

the first nonholonomic constraint writes

$$\dot{x} \sin(\theta + \phi) - \dot{y} \cos(\theta + \phi) - \dot{\theta}\ell\cos\phi = 0.$$

The normal vector fields to the distribution are

$$N_1 = (\sin(\theta + \phi), \, -\cos(\theta + \phi), \, -\ell\cos\phi, \, 0), \, N_2 = (\sin\theta, \, -\cos\theta, \, 0, \, 0).$$

There are two physical alternatives for controls: (i) We select the tangent vector fields

$$X_1 = (\cos\theta, \, \sin\theta, \, \frac{1}{\ell}\tan\phi, \, 0), \; X_2 = (0, 0, 0, 1),$$

and then the dynamic system $\dot{x} = u^1 X_1 + u^2 X_2$ appears, where u^1 = rear driving and u^2 = steering input - here we have a control singularity at the point $\phi = \pm\frac{\pi}{2}$ (this singularity is related to the choice of local coordinates); (ii) we select the tangent vector fields

$$Y_1 = (\cos\theta\cos\phi, \, \sin\theta\cos\phi, \, \frac{1}{\ell}\sin\phi, \, 0), \; Y_2 = (0, 0, 0, 1),$$

and then a new dynamical system $\dot{x} = u^1 Y_1 + u^2 Y_2$ appears, where u^1 = front driving and u^2 = steering input - in this case we have no singularity.

Figure 6.4 Car.

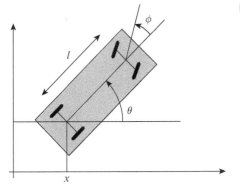

6.5 Nonholonomic Constraints to the *N*-trailer

Let us consider a robot of *N*-trailer type (see Fig. 6.5). We accept that: (i) each trailer is assumed to be connected to the axle midpoint of the previous one (zero hooking); (ii) the generalized coordinates are $(x, y, \phi, \theta_0, \theta_1, ..., \theta_N) \in \mathbb{R}^{N+4}$, where (x, y) is the position of the center of the rear axle, ϕ is the steering angle of the vehicle (in relation to the body of the vehicle), θ_0 is the orientation angle of the vehicle (with respect to Ox axis) and θ_i is the orientation angle of the trailer component i (with respect to Ox axis); (iii) the vehicle is considered as component 0, with the length $d_0 = \ell$, and the length of the component i is d_i (from hook to hook).

The configuration space of the system is the $(n + 3)$-dimensional manifold $Q = SE(2) \times T^n$, where $SE(2)$ denotes the Euclidean group in the plane and T^n is the n-torus.

It follows the nonholonomic constraints (kinematic constraints)

$$\dot{x}_f \sin(\theta_0 + \phi) - \dot{y}_f \cos(\theta_0 + \phi) = 0, \qquad \text{(for front wheels)},$$

$$\dot{x}_i \sin \theta_i - \dot{y}_i \cos \theta_i = 0, \; i = 0, 1, ..., N. \qquad \text{(for all other wheels)}$$

Fixing

$$x_f = x + \ell \cos \theta_0, \; y_f = y + \ell \sin \theta_0,$$

$$x_i = x - \sum_{j=1}^{i} d_j \cos \theta_j, \; y_i = y - \sum_{j=1}^{i} d_j \sin \theta_j,$$

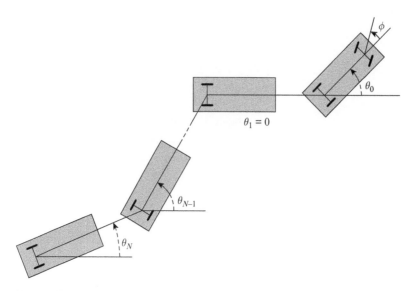

Figure 6.5 N - Trailer.

the constraints become

$$\dot{x} \sin(\theta_0 + \phi) - \dot{y} \cos(\theta_0 + \phi) - \dot{\theta}_0 \ell \cos \phi = 0,$$

$$\dot{x} \sin \theta_i - \dot{y} \cos \theta_i + \sum_{j=1}^{i} \dot{\theta}_j d_j \cos(\theta_i - \theta_j) = 0, \ i = \overline{0, N}.$$

The normal vector fields to the distribution are

$$\xi_{-1} = (\sin(\theta_0 + \phi), \ -\cos(\theta_0 + \phi), \ -\ell \cos \phi, \ 0, \ ..., \ 0),$$

$$\xi_0 = (\sin \theta_0, \ -\cos \theta_0, \ 0, \ ..., \ 0),$$

$$\xi_i = (\sin \theta_i, \ -\cos \theta_i, \ 0, \ d_1 \cos(\theta_i - \theta_1), \ ..., \ d_i \cos(\theta_i - \theta_i), \ 0, \ ..., \ 0),$$

$i = \overline{1, N}.$

We select two vector fields tangent to the distribution:

$$X_1 = (X_1^1, ..., X_1^{N+4}), \tag{1}$$

of components

$$X_1^1 = \cos \theta_0, \ X_1^2 = \sin \theta_0, \ X_1^3 = -\frac{1}{\ell} \tan \phi, \ X_1^4 = -\frac{1}{d_1} \sin(\theta_1 - \theta_0),$$

$$X_1^5 = -\frac{1}{d_2} \cos(\theta_1 - \theta_0) \sin(\theta_2 - \theta_1),$$

$$..., X_1^{i+3} = -\frac{1}{d_i} \left(\prod_{j=1}^{i-1} \cos(\theta_j - \theta_{j-1}) \right) \sin(\theta_i - \theta_{i-1}),$$

$$..., X_1^{N+3} = -\frac{1}{d_N} \left(\prod_{j=1}^{N-1} \cos(\theta_j - \theta_{j-1}) \right) \sin(\theta_N - \theta_{N-1}), \ X_1^{N+4} = 0,$$

and

$$X_2 = (X_2^1, ..., X_2^{N+4}) \tag{2}$$

of components

$$X_2^1 = 0, \ X_2^2 = 0, \ ..., \ X_2^{i+3} = 0, \ ..., \ X_2^{N+3} = 0, \ X_2^{N+4} = 1.$$

It follows the first-order dynamical system $\dot{x} = u^1 X_1 + u^2 X_2$, where $u^1 = $ driving back and $u^2 = $ turn input. Another variant of this evolution is

$$\dot{\theta}_i = -\frac{1}{d_i} v_{i-1} \sin(\theta_i - \theta_{i-1}), \ v_i = v_{i-1} \cos(\theta_i - \theta_{i-1}), \ i = \overline{1, N},$$

where v_i is the linear forward speed of the trailer i and $v_0 = u_1$.

6.6 Famous Lagrangians

(i) The movement of the rolling coin is described by the Lagrangian

$$L = \frac{1}{2}m(\dot{x}^2 + \dot{y}^2) + \frac{1}{2}I_v\dot{\theta}^2 + \frac{1}{2}I_h\dot{\phi}^2,$$

with nonholonomic constraints $\dot{x} = r\dot{\phi}\cos\theta$, $\dot{y} = r\dot{\phi}\sin\theta$.

(ii) The movement of a knife blade is described by the Lagrangian

$$L = \frac{1}{2}m(\dot{x}^2 + \dot{y}^2) + \frac{1}{2}I_v\dot{\theta}^2 + mgy\sin\alpha,$$

with nonholonomic constraint $\dot{x}\sin\theta - \dot{y}\cos\theta = 0$.

(iii) The movement of the Chaplygin skate is described by the Lagrangian

$$L = \frac{1}{2}m(\dot{x}_c^2 + \dot{y}_c^2) + \frac{1}{2}I_v\dot{\theta}^2,$$

with nonholonomic restriction $\dot{x}\sin\theta - \dot{y}\cos\theta = 0$.

(iv) The motion of the Heisenberg system is described by the Lagrangian

$$L = \frac{1}{2}m(\dot{x}^2 + \dot{y}^2 + \dot{z}^2),$$

with nonholonomic constraint $\dot{z} = y\dot{x} - x\dot{y}$.

6.7 Significant Problems

Problem 6.7.1: Let us consider a nonholonomic system on $\mathbb{R}^3$ given by the nonholonomic constraint (ODE with two unknown functions, x and y)

$$\dot{x} + t\dot{y} - y + t = 0.$$

Suppose that the Lagrangian of the system is $L = \frac{1}{2}(\dot{x}^2 + \dot{y}^2)$. Find the geodesics.

Solution. Denote

$$\mathcal{L} = \frac{1}{2}(\dot{x}^2 + \dot{y}^2) + p(t)(\dot{x} + t\dot{y} - y + t).$$

Since

$$\frac{\partial\mathcal{L}}{\partial x} = 0, \frac{\partial\mathcal{L}}{\partial\dot{x}} = \dot{x} + p; \frac{\partial\mathcal{L}}{\partial y} = -p, \frac{\partial\mathcal{L}}{\partial\dot{y}} = \dot{y} + pt,$$

the differential equations of motion are

$$\ddot{x} + \dot{p} = 0, \ \ddot{y} + t\dot{p} = 0, \ \dot{x} + t\dot{y} - y + t = 0.$$

Deriving the nonholonomic constraint and eliminating $\ddot{x}$ and also $\ddot{y}$, we find $(1+t^2)\dot{p} = 1$, with the solution $p(t) = \arctan t + c_1$. It follows

$$\ddot{x} = -\frac{1}{1+t^2}, \ \ddot{y} = -\frac{t}{1+t^2},$$

with the general solution

$$x(t) = -\int (\arctan t + c_1)dt + c_2, \quad y(t) = -\int \left(\frac{1}{2}\ln(1 + t^2) + c_3\right)dt + c_4$$

or

$$x(t) = t(c_1 + \arctan t) - \frac{1}{2}\ln(1 + t^2) + c_2,$$

$$y(t) = -t\left(c_3 - 1 + \frac{1}{2}\ln(1 + t^2)\right) - \arctan t + c_4.$$

Remark 6.1: (i) The nonholonomic constraint $\dot{x} + t\dot{y} - y + t = 0$ is a Pfaff equation on $\mathbb{R}^3$. Indeed, we can write $dx + tdy + (t - y)dt = 0$.

(ii) The inequality $\dot{x} + t\dot{y} - y + t \geq 0$ can be manipulated via a slack variable (a variable that is added to an inequality constraint to transform it into an equality) $\dot{x}(t) + t\dot{y}(t) - y(t) + t - \dot{\xi}^2(t) = 0$.

Problem 6.7.2: Show that the constraint

$$(2x + \sin y - ye^{-x})\frac{dx}{dt} + (x\cos y + \cos y + e^{-x})\frac{dy}{dt} = 0$$

is holonomic (concrete case of the remark that any Pfaff equation in two variables is a holonomic constraint).

Solution. Denote

$$a(x, y) = 2x + \sin y - ye^{-x}, \quad b(x, y) = x\cos y + \cos y + e^{-x}.$$

Since $\frac{\partial a}{\partial y} = \frac{\partial b}{\partial x} = \cos y - e^{-x}$, there exists $\phi(x, y)$ such that $d\phi = adx + bdy$, i.e., $\phi(x, y) = \int_{\Gamma} a(x, y)dx + b(x, y)dy$ (path-independent curvilinear integral), where Γ is a given C^1 curve that joins the points (x_0, y_0) and (x, y). We find $\phi(x, y) = x^2 + x\sin x + ye^{-x} + \sin y$, and the constraint is equivalent to

$$x^2 + x\sin x + ye^{-x} + \sin y = c.$$

Problem 6.7.3: Show that the constraint (Pfaff equation)

$$\dot{x}\sin\theta - \dot{y}\cos\theta = 0$$

is nonholonomic.

Solution. Since the variables are (x, y, θ), we rewrite

$$\dot{x}\sin\theta - \dot{y}\cos\theta + \dot{\theta} \cdot 0 = 0.$$

Introducing the vector field

$$X = (a, b, c), \quad a(x, y, \theta) = \sin\theta, \quad b(x, y, \theta) = \cos\theta, \quad c(x, y, \theta) = 0,$$

the complete integrability condition $(X, \text{rot}\,X) = 0$ is not satisfied.

Problem 6.7.4:   Show that each of the constraints $dx^2 - x^3 dx^1 = 0$ and $dx^3 - x^1 dx^2 = 0$ is nonholonomic on $\mathbb{R}^3$. But if we take them together, what do they represent?

Solution.   For the first part, the technique from Problem 6.7.3 is applied. For the second part we notice that the system

$$dx^2 - x^3 dx^1 = 0, \ dx^3 - x^1 dx^2 = 0$$

leads to $dx^3 - x^1 (x^3 dx^1) = 0$. It follows the general solution

$$x^3 = k e^{\frac{1}{2} x^{1^2}}, \ x^2 = \int k e^{\frac{1}{2} x^{1^2}} dx^1 + c.$$

Thus the system represents an holonomic constraint. Specifically, the system is equivalent to

$$\frac{dx^1}{1} = \frac{dx^2}{x^3} = \frac{dx^3}{x^1 x^3},$$

representing the field lines of the vector field $X = (1, x^3, x^1 x^3)$.

Remark 6.2:   The constraints $(x^2 + y^2)dx + xzdz = 0$ and $(x^2 + y^2)dy + yzdz = 0$ are not integrable separately, but together they can be integrated to

$$x^2 + y^2 + z^2 = c_1, \ \frac{x}{y} = c_2$$

(family of circles).

Problem 6.7.5:   A relativistic particle in space-time $\mathbb{R}^4$ with Minkowski metric can be considered as mechanical system subject to one nonholonomic constraint

$$-(\dot{x}^1)^2 - (\dot{x}^2)^2 - (\dot{x}^3)^2 + (\dot{x}^4)^2 = 0,$$

which is simple affine of degree 2 in velocities. Find Euler–Lagrange ODEs.

Remark 6.3:   The unilateral nonholonomic constraint (inequality)

$$-(\dot{x}^1)^2 - (\dot{x}^2)^2 - (\dot{x}^3)^2 + (\dot{x}^4)^2 \geq 0$$

can be manipulated via a slack variable

$$-(\dot{x}^1)^2 - (\dot{x}^2)^2 - (\dot{x}^3)^2 + (\dot{x}^4)^2 - \dot{\xi}^2 = 0.$$

Problem 6.7.6:   Consider a particle of mass m in a homogeneous gravitational field. The motion of the particle is now subject to a nonholonomic condition $b^2((\dot{x}^1)^2 + (\dot{x}^2)^2)) - (\dot{x}^3)^2 = 0$ (a skleronomic nonholonomic constraint, which is affine of degree 2 in components of velocity), where b is a constant. Find Euler–Lagrange moving equations.

Problem 6.7.7: **(A homogeneous ball on a rotating table)** Consider a homogeneous ball of radius R rolling without sliding on a horizontal plane which rotates with a nonconstant angular velocity $\Omega(t)$ around the vertical axis. We assume that except the constant gravitational force, no other external forces act on the ball.

The potential energy is constant, so without loss of generality we put $V = 0$. In addition, since we do not consider external forces, the Lagrangian is given by the kinetic energy of the rotating ball

$$L = T = \frac{1}{2}\left(\dot{x}^2 + \dot{y}^2 + k^2(\dot{\vartheta}^2 + \dot{\varphi}^2 + \dot{\psi}^2 + 2\dot{\varphi}\,\dot{\psi}\cos\vartheta)\right),$$

where k is the radius of gyration and the mass of the ball is $m = 1$. Find Euler–Lagrange equations of movement.

6.8 Maple Application Topics

Problem 6.8.1: Using EulerLagrange command in Maple, computes the Euler–Lagrange equations of the functional in rolling coin Problem.

Solution. The answer is nonholonomic geodesics.
> with(VariationalCalculus)

$$> L := \frac{1}{2}m((diff(x(t),t))^2 + (diff(y(t),t))^2)$$
$$+ \frac{1}{2}(I_v(diff(\theta(t),t))^2 + I_h(diff(\phi(t),t))^2)$$
$$+ p(t)(r * diff(\phi(t),t)\cos\theta - diff(x(t),t))$$
$$+ q(t)(r * diff(\phi(t),t)\sin\theta - diff(y(t),t)).$$

> EulerLagrange$(L,t,[x(t),y(t),p(t),q(t)])$

Problem 6.8.2: Using EulerLagrange command in Maple, computes the Euler–Lagrange equations of the functional in Problem 6.8.1.

Solution. The answer is nonholonomic geodesics.
> with(VariationalCalculus)

$$> L := \frac{1}{2}((diff(x(t),t))^2 + (diff(y(t),t))^2)$$
$$+ p(t)(diff(x(t),t) + t\,diff(y(t),t) - y + t).$$

> EulerLagrange$(L,t,[x(t),y(t),p(t)])$

Problem 6.8.3: DEtools[*hamilton_eqs*] – generate Hamilton equations

The Hénon–Heiles Hamiltonian

>H := (1/2) * ((p_1)² + (p_2)² + (x^1)² + (x^2)²) + (x^1)² * x^2 – (1/3) * (x^2)³;
> *hamilton_eqs*(H);

Problem 6.8.4: DEtools[*hamilton_eqs*] – generate Hamilton equations

The Hamiltonian $H = \frac{1}{2}(p_1^2 + p_2^2) + \ln(q_1^2 + q_2^2 + 1) + q_2$.

> *hamilton_eqs*(H);

Problem 6.8.5: DEtools[*hamilton_eqs*] – generate Hamilton equations

The Hamiltonian $H = \frac{1}{2}(p_1 + p_2 + Ax^2 + By^2) - \epsilon xy^2$.

> *hamilton_eqs*(H);

Problem 6.8.6: (Vibrating membrane) The Lagrangian of a vibrating membrane is $L = \frac{1}{2}\left(\rho u_t^2 - \tau\left(u_x^2 + u_y^2\right)\right)$. Find the Hamiltonian H, write the discrete version of H and realize a numerical study of the motion.

7

Problems: Free and Constrained Extremals

Motto:
"Who's inclined to mourning, who would like to wail,
Ought to hear their urging – though it's far from plain –
Watch the torch of poplars that from heavens hail,
Burying his shadow where they shade the plain."
Tudor Arghezi – *Never Yet Had Autumn...*

Variational Calculus Problems are intended for circulation concepts and familiarization with the formulas encountered in solving extrema problems on functional spaces. Some mimic the theory, and others increase the ability to calculate and reason. We do not rule out that some problems may appear for the first time in a text addressed to students and researchers. We believe that the given indications and solutions are sufficiently fluidized, allowing the acceleration of user metamorphoses from ignorant to knowledgeable in variational calculus. The problems were formulated by the authors according to the models in the papers [1]–[83].

7.1 Simple Integral Functionals

Problem 7.1.1: Find the Euler–Lagrange ODEs for functionals:

(a) $\int_a^b [u'(x)^2 + e^{u(x)}]\,dx$; (b) $\int_a^b u(x)u'(x)\,dx$; (c) $\int_a^b x^2(u'(x))^2\,dx$.

Problem 7.1.2: The principle of maximum entropy selects the probability distribution that maximizes $H = -\int_a^b u \log u\,dx$. Introduce Lagrange multipliers for the constraints $\int_a^b u\,dx = 1$ and $\int_a^b xu(x)\,dx = \dfrac{1}{a}$, and find by differentiation an Euler–Lagrange equation for $u(x)$. Look for solutions.

Solution. The exact solution is $u(x) = \left|\dfrac{\lambda}{e^\lambda - 1}\right| e^{\lambda x}$, where λ is the unique solution of the equation

Variational Calculus with Engineering Applications, First Edition. Constantin Udriste and Ionel Tevy.
© 2023 John Wiley & Sons Ltd. Published 2023 by John Wiley & Sons Ltd.

$$\frac{e^\lambda}{e^\lambda - 1} - \frac{1}{\lambda} = \frac{1}{a}, \quad a > 1.$$

Problem 7.1.3: Compute the first variation of the functionals:

a) $I(y(\cdot)) = \int_a^b y(x)y'(x)dx;$

b) $I(y(\cdot)) = e^{y(a)};$

c) $I(y(\cdot)) = y^2(0) + \int_0^1 (3y^2(x) + x)dx, \, y_0(x) = x, \, h(x) = x + 1.$

Solution. In all cases we accept the variation $\bar{y}(x) = y(x) + \varepsilon h(x)$.

a) In this case we have

$$I(\varepsilon) = \int_a^b (y(x) + \varepsilon h(x))(y'(x) + \varepsilon h'(x))dx$$

and hence the first variation is the real number

$$\left.\frac{dI}{d\varepsilon}\right|_{\varepsilon=0} = \int_a^b (y(x)h(x))'dx = y(b)h(b) - y(a)h(a).$$

b) The function

$$I(\varepsilon) = e^{y(a)+\varepsilon h(a)}$$

leads to the first variation

$$\left.\frac{dI}{d\varepsilon}\right|_{\varepsilon=0} = h(a)e^{y(a)}.$$

c) Generally, we can write

$$I(\varepsilon) = (y(0) + \varepsilon h(0))^2 + \int_0^1 (3(y(x) + \varepsilon h(x))^2 + x)dx.$$

We find the first variation

$$\left.\frac{dI}{d\varepsilon}\right|_{\varepsilon=0} = 2h(0)y(0) + 6\int_0^1 y(x)h(x)dx.$$

On the other hand, here we have $h(x) = x + 1, \, y(x) = y_0(x) = x$. So

$$\left.\frac{dI}{d\varepsilon}\right|_{\varepsilon=0} = \int_0^1 (6x^2 + 6x)dx = 5.$$

Problem 7.1.4: Find the extremals of functionals:

a) $I(y(\cdot)) = \int_1^2 \frac{1}{x}\sqrt{1 + y'^2(x)} \, dx, \, y(1) = 0, y(2) = 1;$

b) $I(y(\cdot)) = \int_1^2 (y(x)y'(x) + y''^2(x)) \, dx, \, y(0) = 0, \, y'(0) = 1, y(1) = 2, y'(1) = 4;$

c) $I(x(\cdot)) = \displaystyle\int_0^1 [2x(t)^3 + 3t^2 x'(t)]\,dt$, where $x \in C^1[0,1]$, with $x(0) = 0$ and $x(1) = 1$.

What happens if $x(0) = 0$ and $x(1) = 2$?

Solution. In the first problem we use the second-order Euler–Lagrange equation

$$\frac{\partial L_1}{\partial y} - \frac{d}{dx}\frac{\partial L_1}{\partial y'} = 0,$$

and for the second problem we use the fourth-order Euler–Lagrange equation

$$\frac{\partial L_2}{\partial y} - \frac{d}{dx}\frac{\partial L_2}{\partial y'} + \frac{d^2}{dx^2}\frac{\partial L_2}{\partial y''} = 0,$$

whose solutions are called *extremals*.

a) Since the Lagrangian has the expression $L_1(x,y,y') = \dfrac{\sqrt{1+y'^2(x)}}{x}$, we find $\dfrac{\partial L_1}{\partial y} = 0$,

$\dfrac{\partial L_1}{\partial y'} = \dfrac{y'}{x\sqrt{1+y'^2}}$. The Euler–Lagrange equation

$$\frac{y'}{\sqrt{1+y'^2}} = c_1 x$$

rewrites

$$y' = \pm\frac{c_1 x}{\sqrt{1-c_1^2 x^2}}.$$

The general solution is

$$y(x) = \pm\left(\frac{1}{c_1}\sqrt{1-c_1^2 x^2} + c_2\right).$$

Setting the boundary conditions, we find $c_1 = -\dfrac{1}{\sqrt{5}}$, $c_2 = 2$ and hence $y = 2 - \sqrt{5 - x^2}$.

Problem 7.1.5: Find the extrema of simple integral functionals:

a) $I(y(\cdot)) = \displaystyle\int_1^4 \frac{1}{x^3}\, y'^3(x)\, dx$, $y(1) = 0$, $y(4) = 1$;

b) $I(y(\cdot)) = \displaystyle\int_a^b (y^2(x) + y'^2(x) + 2y(x)e^x)\, dx$.

Solution. First we have to find the extremals.

a) Here the Lagrangian is $L(x,y,y') = \dfrac{y'^3}{x^3}$. Since $\dfrac{\partial L}{\partial y} = 0$, $\dfrac{\partial L}{\partial y'} = 3\dfrac{y'^2}{x^3}$, the Euler–Lagrange

ODE is $y'^2 = c_1 x^3$ or $y' = \pm k_1 x^{\frac{3}{2}}$. It follows the general solution $y(x) = \pm k_2 x^{\frac{5}{2}} + k_3$,

where $k_2 > 0$. From the conditions at the ends we obtain the solutions $y_1(x) = \dfrac{1}{31}(x^{\frac{5}{2}} - 1)$

and $y_2(x) = -\dfrac{1}{31}(x^{\frac{5}{2}} + 1)$. Since $\dfrac{\partial^2 L}{\partial y'^2} = \dfrac{6y'}{x^3}$, the Legendre-Jacobi test shows that the

function $y_1(x)$ gives a minimum, and the function $y_2(x)$ gives a maximum.

Problem 7.1.6: Find extremals for the simple integral functional

$$I(x(\cdot)) = \frac{1}{2}\int_0^1 [x'(t)^2 + x(t)x'(t) + x'(t) + x(t)]\, dt,$$

when the values of the function x are free at the endpoints.

Hint. $\frac{\partial L}{\partial x'}(t, x_0(t), x_0'(t)) = 0$ at the free endpoint.

Problem 7.1.7: Minimize the simple integral functional

$$I(y(\cdot)) = \int_0^\infty \left(y^2(x) + y'^2(x) + (y''(x) + y'(x))^2\right) dx,$$

$y(0) = 1, y'(0) = 2, y(\infty) = 0, y'(\infty) = 0.$

Problem 7.1.8: On the vector space $C^1[a, b]$, we consider the following three functionals:

a) $I(y(\cdot)) = \int_a^b y(x)y'(x)dx$; b) $I(y(\cdot)) = e^{y(a)}$; c) $I(y(\cdot)) = \int_a^b (xy'(x) + y(x))dx$. Are any of them linear?

Solution. A functional $I(y(\cdot))$ is called linear if

$$I(\alpha_1 y_1(\cdot) + \alpha_2 y_2(\cdot)) = \alpha_1 I(y_1(\cdot)) + \alpha_2 I(y_2(\cdot)),$$

for any scalars α_1 and α_2. The first two functionals are not linear; the third is linear.

Problem 7.1.9: Minimize the functionals:

a) $I(y(\cdot)) = \int_0^1 (1 + y'^2(x))dx$, $y(0) = 0, y'(0) = 1$;

b) $I(y(\cdot)) = 2\pi \int_0^1 y(x)\sqrt{1 + y'^2(x)}\, dx$, $y(1) = 1, y'(1) = 1$.

Problem 7.1.10: Determine the C^2 function $y(\cdot)$ which minimizes the functional

$$I(y(\cdot)) = (y(1))^2 + \int_0^1 y'^2(x)dx, \, y(0) = 1, y'(0) = 0.$$

Problem 7.1.11: For $x \in D = \{x \in C^1[0, T] \mid x(0) = 0, x(T) = b\}$, we consider the functional

$$I(x(\cdot)) = \int_0^T (c_1\dot{x}^2(t) + c_2x(t))dt.$$

Find the extremals when $T = 1$. Find the extremals for T free.

Problem 7.1.12: Find the extremals of the functional

$$I(x(\cdot)) = \int_0^T \left(\frac{1}{2} \dot{x}^2(t) + x(t)(\dot{x}(t) + 1) \right) dt$$

for $x \in C^2[0, T]$.

Problem 7.1.13: **(Poincaré inequality)** Prove the one-dimensional inequality

$$\int_0^T \phi(t)^2 \, dt \leq \frac{1}{2} T^2 \int_0^T |\dot{\phi}(t)|^2 \, dt,$$

for all C^2 functions ϕ satisfying $\phi(0) = \phi(T) = 0$.

7.2 Curvilinear Integral Functionals

Problem 7.2.1: Let $\Omega = [0, 1]^2 \subset \mathbb{R}^2$ and Γ be an increasing curve in Ω which joins the points $(0, 0)$ and $(1, 1)$. Determine the first variation of the functional (path-independent curvilinear integral)

$$I(u(\cdot)) = \int_\Gamma u_{t^1}(t^1, t^2) \, dt^1 + u_{t^2}(t^1, t^2) \, dt^2,$$

with the boundary conditions $u(0, 0) = 0, u(1, 1) = 1$.

Solution. The path independence conditions are satisfied automatically (the domain is simply connected and the integrand is the differential of the function u). The variation $\bar{u}(t) = u(t) + \varepsilon h(t)$ produces an integral with one parameter

$$I(\varepsilon) = \int_\Gamma (u_{t^1} + \varepsilon h_{t^1}) dt^1 + (u_{t^2} + \varepsilon h_{t^2}) dt^2.$$

The first-order variation is

$$\frac{dI}{d\varepsilon}\bigg|_{\varepsilon=0} = \int_\Gamma h_{t^1} dt^1 + h_{t^2} dt^2 = h(1, 1) - h(0, 0).$$

Simpler The relation $I(u(\cdot)) = u(1, 1) - u(0, 0)$ leads to $I(\varepsilon) = u(1, 1) + \varepsilon h(1, 1) - u(0, 0) - \varepsilon h(0, 0)$.

Problem 7.2.2: Let $\Omega = [0, 1]^2 \subset \mathbb{R}^2$ and Γ be an increasing curve in Ω which joins the points $(0, 0)$ and $(1, 1)$. Find the equations of the extremals of the functional

$$I(u(\cdot)) = \int_\Gamma \left(u(t^1, t^2) + u_{t^2}^2(t^1, t^2) \right) d(t^1 t^2),$$

provided that the curvilinear integral be path-independent.

Problem 7.2.3: Let $\Omega = [0,1]^2$ and Γ be an increasing curve in Ω joining the points $(0,0)$ and $(1,1)$. Determine the first variation of the functional

$$I(u(\cdot)) = \int_\Gamma \left(u + u_{t^1}(t^1, t^2)\right) dt^1 + \left(u + u_{t^2}(t^1, t^2)\right) dt^2,$$

with the boundary conditions $u(0,0) = 0$, $u(1,1) = 1$.

Problem 7.2.4: Let $\Omega = [0,1]^2$ and Γ be an increasing curve in Ω joining the points $(0,0)$ and $(1,1)$. Find the extremals of the functional

$$I(u(\cdot)) = \int_\Gamma \left(u(t^1, t^2) + u_{t^1}^2(t^1, t^2) + u_{t^2}^2(t^1, t^2)\right) d(t^1 t^2),$$

path-independent curvilinear integral, with the boundary conditions $u(0,0) = 0$, $u(1,1) = 1$.

Problem 7.2.5: There is a curve $\Gamma \subset \Omega = [0,1]^3 \subset \mathbb{R}^3$ joining the points $(0,0,0), (1,1,1)$ and which extremizes the functional

$$I(\Gamma) = \int_\Gamma P(x,y,z)dx + Q(x,y,z)dy + R(x,y,z)dz ?$$

Solution. Let us find necessary conditions. Introducing the vector field $F = (P,Q,R)$ and the C^1 curve $\Gamma : x = x(t), y = y(t), z = z(t), t \in [a,b]$, the integral writes

$$I(\Gamma) = \int_a^b (F, \dot{\Gamma}) dt.$$

Suppose that Γ is a curve which extremizes the functional and $\Gamma_\varepsilon : x = x(t,\varepsilon), y = y(t,\varepsilon), z = z(t,\varepsilon), t \in [a,b], \varepsilon \in (-\delta,\delta)$ is a differentiable variation of Γ (surface). We introduce the variation vector field

$$\xi = \frac{\partial \Gamma_\varepsilon}{\partial \varepsilon}\Big|_{\varepsilon=0} : u(t) = \frac{\partial x}{\partial \varepsilon}(t;0), v(t) = \frac{\partial x}{\partial \varepsilon}(t;0), w(t) = \frac{\partial x}{\partial \varepsilon}(t;0),$$

which cancels at the ends. It follows the function (integral with parameter)

$$I(\varepsilon) = \int_a^b (F(x(t,\varepsilon), y(t,\varepsilon), z(t,\varepsilon), \dot{\Gamma}_\varepsilon) dt.$$

A necessary condition is

$$0 = I'(0) = \int_a^b \left((\nabla_\xi F, \dot{\Gamma}) + (F, \dot{\xi})\right) dt.$$

Integrating by parts (last integral), we find

$$0 = I'(0) = \int_a^b \left((\nabla_\xi F, \dot{\Gamma}) - (\nabla_{\dot{\Gamma}} F, \xi)\right) dt.$$

Since the variation vector field ξ is arbitrary, it follows the necessary condition $\langle \nabla F, \dot{\Gamma} \rangle = \nabla_{\dot{\Gamma}} F$. This is an encrypted condition.

Simpler The functional transcribes as a simple integral

$$I(x(\cdot), y(\cdot), z(\cdot)) = \int_a^b (P(x,y,z)\dot{x} + Q(x,y,z)\dot{y} + R(x,y,z)\dot{z})\,dt.$$

The Lagrangian

$$L = P(x,y,z)\dot{x} + Q(x,y,z)\dot{y} + R(x,y,z)\dot{z}$$

has the partial derivatives

$$\frac{\partial L}{\partial x} = \frac{\partial P}{\partial x}\dot{x} + \frac{\partial Q}{\partial x}\dot{y} + \frac{\partial R}{\partial x}\dot{z}, \quad \frac{\partial L}{\partial \dot{x}} = P$$

and similar for the other variables y and z. The first Euler–Lagrange equation is

$$\left(\frac{\partial Q}{\partial x} - \frac{\partial P}{\partial y}\right)\dot{y} + \left(\frac{\partial R}{\partial x} - \frac{\partial P}{\partial z}\right)\dot{z} = 0$$

and the second, the third Euler–Lagrange equations are obtained by cyclic permutations of x, y and z. We find rot $F \times \dot{\Gamma} = 0$, i.e, Γ must be a vortex line (field line of rotor vector field of F).

7.3 Multiple Integral Functionals

Problem 7.3.1: Compute the first variation of double integral functionals:

a) $I(y(\cdot)) = \displaystyle\iint_{[0,1]^2} y(x)y(t)\,\sin(xt)\,dxdt,$

where $y : [0,1] \to \mathbb{R}$ is a function of class C^1;

b) $I(u(\cdot)) = \displaystyle\iint_\Omega (u_x^2(x,y) + u_y^2(x,y)e^{u(x,y)})\,dxdy,$

where Ω is a closed region in the plane xOy and $u : \Omega \to \mathbb{R}$ is a function of class C^1.

Solution. a) We accept the variation $\bar{y}(x) = y(x) + \varepsilon h(x)$. It follows (double integral with one parameter)

$$I(\varepsilon) = \iint_{[0,1]^2} \sin(xt)(y(x) + \varepsilon h(x))(y(t) + \varepsilon h(t))dxdt.$$

Deriving under the integral sign, setting $\varepsilon = 0$, we find the first variation

$$\frac{dI}{d\varepsilon}\Big|_{\varepsilon=0} = \iint_{[0,1]^2} \sin(xt)(y(x)h(t) + y(t)h(x))\,dxdt.$$

b) We accept the variation $\bar{u}(x,y) = u(x,y) + \varepsilon h(x,y)$. It follows (double integral with one parameter)

$$I(\varepsilon) = \iint_\Omega ((u_x + \varepsilon h_x)^2 + (u_y + \varepsilon h_y)^2 e^{u+\varepsilon h})\,dxdy.$$

Deriving under integral sign, setting $\varepsilon = 0$, we deduce the first variation

$$\frac{dI}{d\varepsilon}\Big|_{\varepsilon=0} = \iint_{\Omega} (2u_x h_x + 2u_y h_y e^u + u_y h e^u)\, dxdy.$$

Problem 7.3.2: Let $\Omega = [a, b] \times [a, b]$. Find a necessary condition that the function $y_0(\cdot)$ is a minimum point for the functional

$$I(y(\cdot)) = \iint_{\Omega} K(s, t) y(s) y(t)\, dsdt + \int_a^b y^2(s)\, ds - 2 \int_a^b y(s) f(s)\, ds.$$

Solution. Suppose that the function $y_0(x)$ is a minimum point. We build the variation $\bar{y}(x) = y_0(x) + \varepsilon h(x)$. The function

$$I(\varepsilon) = \iint_{\Omega} K(s, t)(y_0(s) + \varepsilon h(s))(y_0(t) + \varepsilon h(t))\, dsdt$$

$$+ \int_a^b (y_0 + \varepsilon h)^2(s)\, ds - 2 \int_a^b (y_0(s) + \varepsilon h(s)) f(s)\, ds$$

must have $\varepsilon = 0$ as minimum point. It follows the necessary condition

$$0 = \frac{dI}{d\varepsilon}\Big|_{\varepsilon=0} = \iint_{\Omega} K(s, t)(y_0(s) h(t) + y_0(t) h(s))\, dsdt$$

$$+ 2 \int_a^b h(s)(y_0(s) - f(s))\, ds.$$

Otherwise written

$$0 = \frac{dI}{d\varepsilon}\Big|_{\varepsilon=0} = \iint_{\Omega} (K(s, t) + K(t, s)) y_0(t) h(s)\, dsdt$$

$$+ 2 \int_a^b h(s)(y_0(s) - f(s))\, ds$$

$$= \int_a^b \left(2(y_0(s) - f(s)) + \int_a^b (K(s, t) + K(t, s)) y_0(t)\, dt \right) h(s)\, ds.$$

Since h is arbitrary, we find

$$2(y_0(s) - f(s)) + \int_a^b (K(s, t) + K(t, s)) y_0(t)\, dt = 0$$

(integral equation with symmetric kernel, Fredholm equation of the second type).

Problem 7.3.3: Let Ω be a closed region in the plane xOy and $u : \Omega \to \mathbb{R}$ be a function of class C^2. Find the Euler–Lagrange PDE associated to the double integral functional

$$I(u(\cdot)) = \iint_{\Omega} (u_x^2(x, y) + u_y^2(x, y) + 2f(x, y) u(x, y))\, dxdy.$$

Problem 7.3.4: Let us consider $\Omega = [0, 1]^2 \subset \mathbb{R}^2$. Find the extremals of each double integral functional

a) $I(u(\cdot)) = \displaystyle\int_\Omega \left(u(t^1, t^2) + u_{t^1}^2(t^1, t^2) + u_{t^2}^2(t^1, t^2) \right) dt^1 dt^2,$

b) $I(u(\cdot)) = \displaystyle\int_\Omega u_{t^1}^2(t^1, t^2) u_{t^2}^2(t^1, t^2) \, dt^1 dt^2,$

c) $I(u(\cdot)) = \displaystyle\int_\Omega \sqrt{1 + u_{t^1}^2(t^1, t^2) + u_{t^2}^2(t^1, t^2)} \, dt^1 dt^2,$

with the boundary conditions $u(0, 0) = 0$, $u(1, 1) = 1$.

Solution. Being given the Lagrangian L, the extremals are solutions of second-order Euler–Lagrange PDE

$$\frac{\partial L}{\partial u} - D_{t^1} \frac{\partial L}{\partial u_{t^1}} - D_{t^2} \frac{\partial L}{\partial u_{t^2}} = 0.$$

a) In this case we have

$$L = u(t^1, t^2) + u_{t^1}^2(t^1, t^2) + u_{t^2}^2(t^1, t^2).$$

We find

$$\frac{\partial L}{\partial u} = 1, \quad \frac{\partial L}{\partial u_{t^1}} = 2u_{t^1}, \quad \frac{\partial L}{\partial u_{t^2}} = 2u_{t^2}.$$

The Euler–Lagrange PDE is reduced to a Poisson PDE

$$1 - 2u_{t^1 t^1} - 2u_{t^2 t^2} = 0.$$

Problem 7.3.5: Is there a parallelepiped $\Omega = [0, a] \times [0, b] \times [0, c]$ which extremizes the triple integral functional

$$I(\Omega) = \int_\Omega f(x, y, z) \, dx dy dz \, ?$$

Solution. Suppose the parallelepiped Ω makes the minimum of functional. The deformed parallelepiped $\Omega_{hkl} = [0, a + h] \times [0, b + k] \times [0, c + l]$ leads to the function

$$I(h, k, l) = \int_{\Omega_{hkl}} f(x, y, z) dx dy dz = \int_0^{a+h} \int_0^{b+k} \int_0^{c+l} f(x, y, z) dx dy dz.$$

The necessary conditions of extremum are

$$I_h(0,0,0) = \int_0^b \int_0^c f(a,y,z)dydz = 0,$$

$$I_k(0,0,0) = \int_0^a \int_0^c f(x,b,z)dxdz = 0,$$

$$I_l(0,0,0) = \int_0^a \int_0^b f(x,y,c)dxdy = 0.$$

Analogously, we find

$$I_{hh}(0,0,0) = \int_0^b \int_0^c f_x(a,y,z)dydz, \quad I_{hk}(0,0,0) = \int_0^c f(a,b,z)dz$$

and those obtained by circular permutations. The condition

$$d^2 I(0,0,0) > 0$$

(in positively definite sense) leads to a minimum. The solution to the problem depends on the function f.

Simpler $I(\Omega) = I(a,b,c)$. The necessary conditions to be extremum are $I_a = I_b = I_c = 0$, etc.

Problem 7.3.6: Determine a PDE for the surface defined parametrically by $r : B_1(0) \subset \mathbb{R}^2 \to \mathbb{R}^3$, $(u,v) \mapsto r(u,v) = (x(u,v), y(u,v), z(u,v))$, with $r|_{\partial B_1} = \gamma$ (closed curve), and which minimizes the area

$$\int_{B_1} \sqrt{\det(\langle Dr, Dr \rangle)}\, dudv.$$

7.4 Lagrange Multiplier Details

For abbreviated expression we make up doublets and triplets, the first element being the objective functional, the second and possibly the third representing the constraints.

The most simple situations are: the pair (simple integral functional, ODEs) asks for Lagrange multipliers like functions of integration parameter, the pair (simple integral functional, nonholonomic constraints) needs Lagrange multipliers like functions of integration parameter, and the pair (simple integral functional, simple integral functionals (isoperimetric constraints)) requires Lagrange multipliers like constants.

Differential forms are an approach to multivariable calculus that is independent of coordinates. They provide a unified approach to define integrands over curves, surfaces, solids and higher-dimensional manifolds.

On $\mathbb{R}^3$, the curvilinear integral (second type) functionals, surface integral (second type) functionals and triple integral functionals are written with the help of differential Lagrange

1-forms

$$\varphi = L_1(t, x(t), x_\gamma(t)) \, dt^1 + L_2(t, x(t), x_\gamma(t)) \, dt^2 + L_3(t, x(t), x_\gamma(t)) \, dt^3,$$

differential Lagrange 2-forms

$$\psi = L_{23}(t, x(t), x_\gamma(t)) \, dt^2 \wedge dt^3 + L_{31}(t, x(t), x_\gamma(t)) \, dt^3 \wedge dt^1$$
$$+ L_{12}(t, x(t), x_\gamma(t)) \, dt^1 \wedge dt^2,$$

and differential Lagrange 3-forms

$$\omega = L(t, x(t), x_\gamma(t)) \, dt^1 \wedge dt^2 \wedge dt^3,$$

respectively. The 1-forms are transferred in 2-forms, and the 2-forms are transferred in 3-forms by exterior differentiation d.

Each of the pairs (curvilinear integral functional, PDEs; surface integral functional, PDEs; triple integral functional, PDEs) requires Lagrange multipliers like functions of integration parameters. Each of the pairs (curvilinear integral functional, curvilinear integral functional (isoperimetric constraint); surface integral functional, surface integral functional (isoperimetric constraint); triple integral functional, triple integral functional (isoperimetric constraint)) requires constant Lagrange multipliers.

The triple (curvilinear integral functional, surface integral functional (isope- rimetric constraint), triple integral functional (isoperimetric constraint)) imposes a vector field factor $p = p^\alpha \frac{\partial}{\partial t^\alpha}$, $\alpha = \overline{1,3}$, for the first exterior product by which the Lagrange 2-form is transformed into a Lagrange 1-form, and a tensor field factor

$$q = q^{23} \frac{\partial}{\partial t^2} \otimes \frac{\partial}{\partial t^3} + q^{31} \frac{\partial}{\partial t^3} \otimes \frac{\partial}{\partial t^1} + q^{12} \frac{\partial}{\partial t^1} \otimes \frac{\partial}{\partial t^2},$$

for the second interior product by which the Lagrange 3-form is transformed into a Lagrange 1-form, to make possible the linear combination.

The triple (triple integral functional, curvilinear integral functional (isope- rimetric constraint), surface integral functional (isoperimetric constraint)) imposes a 2-form factor ψ for the first exterior product by which the Lagrange 1-form is transformed into a Lagrange 3-form, and an 1-form factor φ for the second exterior product by which the Lagrange 2-form is transformed into a Lagrange 3-form, to make possible the linear combination.

7.5 Simple Integral Functionals with ODE Constraints

Problem 7.5.1: Find the extrema of total curvature

$$K(y(\cdot)) = \int_a^b \frac{y''(x)}{(1 + y'^2(x))^{\frac{3}{2}}} \, dx,$$

among the graphs of the solutions of the Bessel equation.

Solution. The Bessel functions are solutions of ODE

$$x^2\, y''(x) + x\, y'(x) + (x^2 - \nu^2)\, y(x) = 0,\ \nu \in \mathbb{R} \setminus \mathbb{Z}.$$

If ν is not an integer, then the solutions of the Bessel equation have the form

$$y(x) = c_1 J_\nu(x) + c_2 J_{-\nu}(x).$$

The functional is replaced by the function (integral with two parameters)

$$K(c_1, c_2) = \int_a^b \frac{c_1 J_\nu''(x) + c_2 J_{-\nu}''(x)}{(1 + (c_1 J_\nu'(x) + c_2 J_{-\nu}')(x))^2)^{\frac{3}{2}}}\, dx.$$

The extremals correspond to the solutions (c_1, c_2) of the system

$$\frac{\partial K}{\partial c_1} = 0,\ \frac{\partial K}{\partial c_2} = 0.$$

To this system, we must add the Bessel function ODE.

Remark 7.1: (i) Let us confine ourselves to the graphs of Bessel functions that satisfy the relationship $y'^2(x) < 1$. First we observe that

$$K(y(\cdot)) \cong \int_a^b y''(x) \left(1 + \frac{3}{2} y'^2(x)\right) dx = J(y(\cdot)),$$

and we accept that the interval $[a, b]$ does not contain zero. Then we can remove on y'', reducing the functional to

$$J(y(\cdot)) = \int_a^b \left(-\frac{1}{x} y' + \left(\frac{\nu^2}{x^2} - 1\right) y\right) \left(1 + \frac{3}{2} y'^2\right) dx.$$

To solve the problem, we need the Lagrangian

$$L(x, y, y') = \left(-\frac{1}{x} y' + \left(\frac{\nu^2}{x^2} - 1\right) y\right) \left(1 + \frac{3}{2} y'^2\right)$$
$$+ p(x)(x^2 y''(x) + x y'(x) + (x^2 - \nu^2) y(x)).$$

(ii) If we change the ODE into an inequality

$$x^2\, y''(x) + x\, y'(x) + (x^2 - \nu^2)\, y(x) \geq 0,\ \nu \in \mathbb{R} \setminus \mathbb{Z},$$

then we must introduce a slack variable

$$x^2\, y''(x) + x\, y'(x) + (x^2 - \nu^2)\, y(x) - \xi'(x)^2 = 0.$$

Problem 7.5.2: Find Hermite EDO solutions that have minimal total curvature graphs.

Solution. The total curvature of a graph $y = y(x)$ has the expression

$$K(y(\cdot)) = \int_a^b \frac{y''(x)}{(1 + y'^2(x))^{\frac{3}{2}}} \, dx.$$

The Hermite ODE is

$$y''(x) - 2x\, y'(x) + 2n\, y(x) = 0.$$

Eliminating $y''(x)$, we introduce the Lagrangian

$$L(x, y(x), y'(x)) = 2\frac{xy'(x) - ny(x)}{(1 + y'^2(x))^{\frac{3}{2}}} + p(x)(y''(x) - 2xy'(x) + 2ny(x)).$$

The solutions to the problem are among the solutions of a system: Euler–Lagrange equation and the Hermite ODE.

   Removing method: The replacement of y'' from restriction in the original Lagrangian modifies it, but the optimal problem remains as a constrained problem.

Problem 7.5.3: Find solutions to the Hermite equation that have graphs of minimum elastic potential energy.

Solution. The *elastic potential energy* of a graph $y = y(x)$ has the expression (under the integral we recognize the square of the curvature)

$$E(y(\cdot)) = \frac{1}{2} \int_a^b \frac{y''^2(x)}{(1 + y'^2(x))^3} \, dx.$$

On the other hand, the Hermite ODE is

$$y''(x) - 2x\, y'(x) + 2n\, y(x) = 0,\ n \in N, \text{fixed.}$$

It follows the Lagrangian

$$L = \frac{xy'(x) - ny(x)}{(1 + y'^2(x))^3} + p(x)(y''(x) - 2xy'(x) + 2ny(x)).$$

We compute

$$\frac{\partial L}{\partial y} = \frac{-n}{(1 + y'^2)^3} + 2np(x),\quad \frac{\partial L}{\partial y'} = \frac{x - 5xy' + 6nyy'}{(1 + y'^2)^4} - 2xp(x),$$

$$\frac{\partial L}{\partial y''} = p(x),\quad \frac{\partial L}{\partial p} = y''(x) - 2xy'(x) + 2ny(x).$$

We write the Euler–Lagrange ODEs and we look for solutions.

Problem 7.5.4: Let us consider the functional

$$I(y(\cdot)) = y^2(0) + \int_0^1 (3y^2(x) + x)\, dx.$$

Determine the extremals constrained by ODE

$$y'(x) = \sqrt{1 + x + y(x)}.$$

Solution. **The first way** If there exists $x_0 \in [0,1]$ such that $1 + x_0 + y(x_0) = 0$, then $y'(x_0) = 0$. The existence condition $1 + x + y(x) > 0$ imposes $y'(x) > 0$.

First we transcribe the constrained in an equivalent form $y'^2(x) = 1 + x + y(x)$. The Lagrangian

$$L(x, y(x), y'(x), p(x)) = 3y^2(x) + x + p(x)\Big(1 + x + y(x) - y'^2(x)\Big)$$

changes the problem of constrained extremals in the problem of free extremals. Since

$$\frac{\partial L}{\partial y} = 6y + p, \quad \frac{\partial L}{\partial y'} = -2py', \quad \frac{\partial L}{\partial p} = 1 + x + y - y'^2,$$

it follows the ODE system

$$6y + p + 2\frac{d}{dx}(py') = 0, \quad 1 + x + y - y'^2 = 0.$$

Solving the constrained equation, we obtain the extremal.

The second way We are looking for a parametric solution for the restriction equation. Denote $p = y'$, i.e., $p^2 = 1 + x + y$. Deriving with respect to x, we use $p' = \frac{1}{x'}$. It follows the parametric solution

$$x(p) = 2p - 2\ln(p + 1) + c, \quad y(p) = p^2 - 2p - 1 - 2\ln(p + 1) - c.$$

From the condition $x(0) = 0$, we find $c = 0$. Then we determine the number p_0 which is the solution of the equation $x(p_0) = 1$. The function $x : [0, p_0] \to \mathbb{R}$ is strictly increasing. It follows

$$I(y(\cdot)) = 1 + 2\int_0^{p_0} (3y^2(p) + x(p))\frac{p}{1 + p}\, dp.$$

The third way We solve the restriction ODE setting $z = \sqrt{1 + x + y}$. We find the equation with separable variables $1 + z = 2zz'$ with an implicit solution

$$z - \ln(z + 1) = \frac{x}{2} + c.$$

The ODE restriction is changed into the algebraic equation

$$\sqrt{1 + x + y} - \ln(1 + \sqrt{1 + x + y}) = \frac{x}{2} + c.$$

Moving everything to the left and denoting $F(x, y, c) = 0$, this equation defines the implicit function $y = \varphi(x, c)$, with the derivative

$$\frac{\partial y}{\partial c} = -\frac{\frac{\partial F}{\partial c}}{\frac{\partial F}{\partial y}} = -2\left(1 + \sqrt{1 + x + y}\right).$$

The functional is changed into the function (integral with one parameter)

$$I(c) = \varphi^2(0, c) + \int_0^1 (3\varphi^2(x, c) + x)dx.$$

The condition

$$I'(c) = 2\varphi(0, c)\varphi'(0, c) - 12\int_0^1 \varphi(x, c)\left(1 + \sqrt{1 + x + \varphi(x, c)}\right) dx = 0$$

fixes the constant c and hence the extremal.

Remark 7.2: The method of removing the restriction does not always lead to a solution. To convince yourself, try the next problem.

Problem 7.5.5: Find the extrema of the simple integral functional

$$J(y(\cdot)) = \int_0^1 \sqrt{1 + y'^2(x)} \, dx,$$

constrained by $y'(x) = \sqrt{1 + y(x)}$.

Solution. The restriction ODE has the general solution

$$y(x) = \frac{1}{4}(x + c)^2 - 1,$$

which replaced into functional gives the real function

$$J(c) = \int_0^1 \sqrt{1 + \frac{1}{4}(x + c)^2} \, dx.$$

The necessary extremum condition is

$$0 = J'(c) = \frac{1}{2}\int_0^1 \frac{x + c}{\sqrt{(x + c)^2 + 4}} \, dx = \frac{1}{2}\left(\sqrt{(1 + c)^2 + 4} - \sqrt{c^2 + 4}\right),$$

whence $c = -\frac{1}{2}$. Since $J''(-\frac{1}{2}) = \frac{2}{\sqrt{17}} > 0$, the function that performs the minimum is

$$y(x) = \frac{1}{4}\left(x - \frac{1}{2}\right)^2 - 1.$$

The minimum value $J(y)$ of the integral is $\frac{1}{8}\sqrt{17} + 4\ln 2 - 2\ln(\sqrt{17} - 1)$.

Removing method: The replacement of y' from restriction in the original Lagrangian $L = \sqrt{1 + y'^2(x)}$ gives $L_1 = \sqrt{2 + y(x)}$, but the problem does not become free of restriction.

Remark 7.3: The integral curves of the constraint equation form a family of parables that are obtained from each other by translation in the direction of the axis Ox. The extremum condition chooses from these parables the one in which the vertical band $[0,1] \times \mathbb{R}$ cuts an arc of minimum length (verify that $J''(-\frac{1}{2}) > 0$).

Problem 7.5.6: Minimize the functional

$$I(y(\cdot)) = \int_1^2 (1 + y'^2(x))dx$$

constrained by Euler ODE

$$x^2 y''(x) - x y'(x) - 3 y(x) = 0, \ x > 0, \ y(1) = 1.$$

Solution. The general solution of the given Euler ODE is $y = c_1 \frac{1}{x} + c_2 x^3$. Setting the condition $y(1) = 1$, we find $c_1 + c_2 = 1$. The functional is transformed into the function

$$I(c_1) = \int_1^2 \left(1 + \left(-\frac{c_1}{x^2} + 3(1 - c_1)x^2\right)^2\right) dx.$$

Since $I''(c_1) > 0$, from $0 = I'(c_1)$, we find the constant c_1 which leads us to the minimizing extremal.

Variant We use the Lagrangian

$$L = 1 + y'^2(x) + p(x)(x^2 y''(x) - xy'(x) - 3y(x)).$$

Since

$$\frac{\partial L}{\partial y} = -3p, \ \frac{\partial L}{\partial y'} = 2y' - xp, \ \frac{\partial L}{\partial y''} = x^2 p, \ \frac{\partial L}{\partial p} = x^2 y''(x) - xy'(x) - 3y(x),$$

it follows the system

$$-3p - \frac{d}{dx}(2y' - xp) + \frac{d^2}{dx^2}(x^2 p) = 0, \ x^2 y''(x) - x y'(x) - 3 y(x) = 0.$$

We must also return to solving the restriction equation.

Problem 7.5.7: For $x \in D = \{x \in C^1[0,T] \mid x(0) = 0, x(T) = b\}$, we consider the functional

$$I(x(\cdot)) = \int_0^T (c_1 \dot{x}^2(t) + c_2 x(t)) dt.$$

Find the extremals constrained by the inequality

$$t \ddot{x}(t) + 2 \dot{x}(t) + t x(t) \geq 1.$$

7.6 Simple Integral Functionals with Nonholonomic Constraints

Problem 7.6.1: Find the curves in the distribution

$$dz = ydx$$

which have the minimum energy

$$I(x(\cdot), y(\cdot), z(\cdot)) = \frac{1}{2} \int_0^1 (\dot{x}^2(t) + \dot{y}^2(t) + \dot{z}^2(t)) \, dt$$

(geodesics).

Solution. The Pfaff equation transcribes as ODE $\dot{z}(t) = y(t)\dot{x}(t)$, $t \in I \subset \mathbb{R}$, with three unknown functions $x(\cdot), y(\cdot), z(\cdot)$. The Lagrangian

$$L = \frac{1}{2}(\dot{x}^2(t) + \dot{y}^2(t) + \dot{z}^2(t)) + p(t)(y(t)\dot{x}(t) - \dot{z}(t))$$

changes the problem of constrained extremals into the problem of free extremals. Since

$$\frac{\partial L}{\partial x} = 0, \; \frac{\partial L}{\partial y} = p\dot{x}, \; \frac{\partial L}{\partial z} = 0, \; \frac{\partial L}{\partial p} = y(t)\dot{x}(t) - \dot{z}(t)$$

$$\frac{\partial L}{\partial \dot{x}} = \dot{x} + py, \; \frac{\partial L}{\partial \dot{y}} = \dot{y}, \; \frac{\partial L}{\partial \dot{z}} = \dot{z} - p$$

it follows the ODE system

$$\frac{d}{dt}(\dot{x} + py) = 0, \; p\dot{x} - \ddot{y} = 0, \; \frac{d}{dt}(\dot{z} - p) = 0, \; y(t)\dot{x}(t) - \dot{z}(t) = 0.$$

This is equivalent to

$$\dot{x} + py = c_1, \; p\dot{x} - \dot{y} = 0, \; \dot{z} - p = c_2, \; y(t)\dot{x}(t) - \dot{z}(t) = 0.$$

The system can be writen

$$p = \frac{c_1 - \dot{x}}{y} = \frac{\ddot{y}}{\dot{x}} = \dot{z} - c_2 = \dot{x}y - c_2,$$

or

$$\frac{c_1 - \dot{x}}{y} = \frac{\ddot{y}}{\dot{x}} = \dot{x}y - c_2.$$

Then it turns out successively $\dot{x} = \dfrac{c_1 + c_2 y}{y^2 + 1}$, $\ddot{y}\dot{y} = (\dot{x}^2 y - c_2 \dot{x})\dot{y}$, and finally

$$\dot{y} = \sqrt{c_3 - \frac{2c_1 c_2 y + c_1^2 - c_2^2}{y^2 + 1}}.$$

Now, theoretically we can find $y(t) = \psi(t, c_1, \dots, c_4)$, then $x(t) = \varphi(t, c_1, \dots, c_5)$ and $z(t) = \chi(t, c_1, \dots, c_6)$. **Another way** We consider a deviated nonholonomic constraint $\dot{z} = \dot{x}y - k_1$ (a holonomic deviation according to t of the initial Pfaff equation) and replace it in

Lagrangian. We obtain the new Lagrangian $\mathcal{L} = \frac{1}{2}(\dot{x}^2 + \dot{y}^2 + (\dot{x}\,y - k_1)^2)$, with E–L equations $\dot{x} + (\dot{x}\,y - k_1)y = k_2$, $\ddot{y} = (\dot{x}\,y - k_1)\dot{x}$. It follows $\dot{x} = \dfrac{k_1 y + k_2}{y^2 + 1}$, and so on as above.

Problem 7.6.2: Using the curvilinear abscissa, determine the curves in distribution

$$dz = \frac{1}{2}(x\,dy - y\,dx)$$

which have extremum total curvature.

Solution. We transform the Pfaff equation into an ODE

$$\dot{z}(s) = \frac{1}{2}(x(s)\dot{y}(s) - y(s)\dot{x}(s)),$$

with three unknown functions $x(s), y(s), z(s)$, where s is the curvilinear abscissa. On the other hand, the total curvature is the functional

$$I(\alpha(\cdot)) = \int_a^b \|\ddot{\alpha}(s)\| ds = \int_a^b \sqrt{\ddot{x}^2(s) + \ddot{y}^2(s) + \ddot{z}^2(s)}\, ds.$$

We introduce the second-order Lagrangian

$$L = \sqrt{\ddot{x}^2(s) + \ddot{y}^2(s) + \ddot{z}^2(s)} + p(s)\left(\frac{1}{2}(x(s)\dot{y}(s) - y(s)\dot{x}(s)) - \dot{z}(s)\right).$$

The curves sought must be among the solutions of the Euler–Lagrange equations produced by L.

Problem 7.6.3: Let $\alpha(t) = (x(t), y(t), z(t))$, $t \in [a, b]$ be a C^2 curve. Find the extrema of the functional $I(\alpha(\cdot)) = \int_a^b \dot{x}(t)\dot{y}(t)\, dt$, with nonholonomic constraint $dz = \frac{1}{2}y^2 dx$.

Solution. We transform the Pfaff equation into ODE

$$\dot{z}(t) = \frac{1}{2}y^2(t)\dot{x}(t),$$

with three unknown functions $x(t), y(t), z(t)$. We use the Lagrangian

$$L = \dot{x}(t)\dot{y}(t) + p(t)\left(\frac{1}{2}y^2(t)\dot{x}(t) - \dot{z}(t)\right).$$

The partial derivatives

$$\frac{\partial L}{\partial x} = 0, \quad \frac{\partial L}{\partial y} = py\dot{x}, \quad \frac{\partial L}{\partial z} = 0, \quad \frac{\partial L}{\partial p} = \frac{1}{2}y^2(t)\dot{x}(t) - \dot{z}(t)$$

$$\frac{\partial L}{\partial \dot{x}} = \dot{y} + \frac{1}{2}py^2, \quad \frac{\partial L}{\partial \dot{y}} = \dot{x}, \quad \frac{\partial L}{\partial \dot{z}} = -p$$

lead to Euler–Lagrange ODEs

$$\dot{y} + \frac{1}{2}py^2 = c_1, \quad py\dot{x} - \ddot{x} = 0, \quad p(t) = 2c_2, \quad \frac{1}{2}y^2(t)\dot{x}(t) - \dot{z}(t) = 0.$$

Let us consider the separable variable equation $\frac{dy}{dt} = c_1 - c_2 y^2$. Since

$$\int \frac{dy}{c_1 - c_2 y^2} = \begin{cases} \frac{1}{\sqrt{-4c_1 c_2}} \ln \frac{-c_2 y - \sqrt{-c_1 c_2}}{-c_2 y + \sqrt{-c_1 c_2}} & \text{for} \quad c_1 c_2 < 0 \\ \frac{1}{\sqrt{c_1 c_2}} \arctan \frac{-c_2 y}{\sqrt{c_1 c_2}} & \text{for} \quad c_1 c_2 > 0, \end{cases}$$

the solution of the differential equation in the unknown y is written immediately. Also, the general solution of the Pfaff equation can be written in the form $x = \int f(t)dt, y = 2t,$

$z = \int 2t^2 f(t)dt$, where f is an arbitrary C^1 function.

Problem 7.6.4: Determine the extremals of the functional

$$I(x(\cdot), y(\cdot), \theta(\cdot)) = \frac{1}{2} \int_0^1 (\dot{x}^2(t) + \dot{y}^2(t) + \dot{\theta}^2(t)) \, dt$$

with nonholonomic constraint $\dot{x} \sin\theta - \dot{y} \cos\theta = 0$ (geodesics).

Problem 7.6.5: Characterize the curves in the distribution

$$\dot{x} + t\dot{y} - y + t = 0$$

which have the minimum length (geodesics).

Solution. The length of a plane curve generates the functional

$$\ell(x(\cdot), y(\cdot)) = \int_a^b \sqrt{\dot{x}^2(t) + \dot{y}^2(t)} \, dt.$$

We introduce the Lagrangian

$$L = \sqrt{\dot{x}^2(t) + \dot{y}^2(t)} + p(t)(\dot{x} + t\dot{y} - y + t).$$

We compute the partial derivatives

$$\frac{\partial L}{\partial x} = 0, \quad \frac{\partial L}{\partial y} = -p,$$

$$\frac{\partial L}{\partial \dot{x}} = \frac{\dot{x}}{\sqrt{\dot{x}^2 + \dot{y}^2}} + p, \quad \frac{\partial L}{\partial \dot{y}} = \frac{\dot{y}}{\sqrt{\dot{x}^2 + \dot{y}^2}} + tp.$$

It follows the Euler–Lagrange ODEs

$$\frac{\dot{x}(t)}{\sqrt{\dot{x}^2(t) + \dot{y}^2(t)}} + p(t) = c_1, \quad p(t) + \frac{d}{dt}\left(\frac{\dot{y}(t)}{\sqrt{\dot{x}^2(t) + \dot{y}^2(t)}} + tp(t)\right) = 0.$$

Remark 7.4: The nonholonomic constraint $\dot{x} + t\dot{y} - y + t = 0$ is a Pfaff equation on $\mathbb{R}^3$. Indeed, we can write $dx + tdy + (t - y)dt = 0$. The general solution of this Pfaff equation is $x = f(y,t), y = f_t - f_y, t = -f_y$, where f is an arbitrary C^2 function.

7.7 Simple Integral Functionals with Isoperimetric Constraints

Problem 7.7.1: Minimize the functional

$$I(y(\cdot)) = \int_0^1 (y'(x)^2 + x^2)\, dx$$

subject to the constraints

$$J(y(\cdot)) = \int_0^1 y(x)^2\, dx = 1,\ y(0) = 1,\ y(1) = 2.$$

Solution. We use the Lagrangian

$$L = y'(x)^2 + x^2 + \lambda y(x)^2,$$

where λ is a constant multiplier. Compute the partial derivatives

$$\frac{\partial L}{\partial y} = 2\lambda y,\ \frac{\partial L}{\partial y'} = 2y'.$$

It follows the Euler–Lagrange ODE $\lambda y(x) - y''(x) = 0$, with the general solution $y(x) = c_1 e^{x\sqrt{\lambda}} + c_2 e^{-x\sqrt{\lambda}}$. From the conditions at the ends and from the restriction, we determine the constants c_1, c_2, λ. The found extremal is a minimum point since $\frac{\partial^2 L}{\partial y'^2} > 0$ (Legendre-Jacobi test).

Problem 7.7.2: Find the extremals of the functional

$$I(y(\cdot)) = \int_0^\pi y'(x)^2\, dx$$

with the conditions

$$J(y(\cdot)) = \int_0^1 y^2(x)dx = 1,\ y(0) = 1,\ y(0) = 0.$$

Problem 7.7.3: Use the Lagrange multipliers method to find candidates for solutions to the problem

$$\min I(x(\cdot)) = \int_0^T e^{-rt} x(t)\, dt$$

with the isoperimetric constraint $x \in D = \left\{ x \in C^1[0,T],\ \int_0^T \sqrt{x(t)}\, dt = A \right\}$, where $r > 0$, $A \geq 0$.

Solution. We use the Lagrangian

$$L = e^{-rt}x(t) + p\sqrt{x(t)},$$

where the multiplier p is a constant. Since $\frac{\partial L}{\partial x} = e^{-rt} + p\,\frac{1}{2\sqrt{x(t)}}, \frac{\partial L}{\partial \dot{x}} = 0$, it follows the Euler–Lagrange equation $e^{-rt} + p\,\frac{1}{2\sqrt{x(t)}} = 0$, with the solution $x(t) = \frac{p^2}{4}e^{-2rt}$. The constant p is determined from the restriction.

Problem 7.7.4: We repeat the question for

$$\min I(x(\cdot)) = \int_0^T \sqrt{1 + x(t)}\,dt$$

and $x \in D = \left\{x \in C^1[0,T], \int_0^T x(t)\,dt = a\right\}$, where $T > 0, a \geq 0$.

Problem 7.7.5: **(Elastic curves)** The potential elastic energy of a C^2 curve $c : [0,T] \to \mathbb{R}^2$ is the functional

$$I(c(\cdot)) = \frac{1}{2}\int_0^T k^2(s)\,ds$$

(the integral of curvature square). A curve of class C^2, which minimizes this functional, is called *elastic curve*. Determine all plane elastic curves of fixed length, passing through two given points and having prescribed speeds at the ends.

Solution. Let $c(s) = (x(s), y(s))$ be a unit speed curve and $\theta(s)$ the angle between velocity $\dot{c}$ and the Ox axis. Then $k^2(s) = \dot{\theta}^2(s)$. Taking into account the speed restriction $\dot{x} = \cos\theta, \dot{y} = \sin\theta$, the problem of finding elastic curves of length T can be solved optimizing the functional

$$J(c(\cdot)) = \int_0^T L(x, y, \theta, \dot{x}, \dot{y}, \dot{\theta})\,ds,$$

where

$$L = \frac{1}{2}\dot{\theta}^2 + \lambda_1(\dot{x} - \cos\theta) + \lambda_2(\dot{y} - \sin\theta).$$

The Euler–Lagrange ODEs

$$\ddot{\theta} = \lambda_1\sin\theta - \lambda_2\cos\theta, \ \dot{\lambda}_1 = 0, \dot{\lambda}_2 = 0$$

lead to the pendulum equation

$$\ddot{u} + a^2\sin\theta = 0,$$

where

$$a^2 = \sqrt{\lambda_1^2 + \lambda_2^2}, \ u = \theta + \pi - \alpha, \ \alpha = \operatorname{arctg}\frac{\lambda_2}{\lambda_1}.$$

The solutions of pendulum equation are written using elliptic functions.

7.8 Multiple Integral Functionals with PDE Constraints

Problem 7.8.1: Let $D : x^2+y^2 < 1$. Verify if the function $u(x,y) = axy(x^2+y^2)$, $(x,y) \in D$, is an extremal of the functional

$$I(u(\cdot)) = \frac{1}{2} \iint_D (u_x^2 + u_y^2 - 48xyu)\, dxdy,$$

with PDE constraint

$$x\frac{\partial u}{\partial x} - y\frac{\partial u}{\partial y} = 2axy(x^2 - y^2).$$

Problem 7.8.2: Determine the extrema of the functional

$$I(z(\cdot)) = \frac{1}{2} \iint_{[0,1]^2} (z_x^2 + z_y^2)\, dxdy,$$

constrained by PDE $z_x z_y - z = 0$.

Solution. A solution of the restriction is $z = (x-a)(y-b)$. Then the functional becomes

$$I(a,b) = \frac{1}{2} \iint_{[0,1]^2} ((x-a)^2 + (y-b)^2)\, dxdy.$$

Since $d^2I > 0$, the conditions $\frac{\partial I}{\partial a} = 0$, $\frac{\partial I}{\partial b} = 0$ give the minimizing extremal.

Problem 7.8.3: Determine the positive functions $g_{11}(x,y)$ and $g_{22}(x,y)$ such that the function $z(x,y) = (x-y)^3 + e^{x+y}$ be an extremal of the functional

$$I(z(\cdot)) = \frac{1}{2} \iint_{[0,1]^2} (g_{11}(x,y)z_x^2 + g_{22}(x,y)z_y^2)\, dxdy.$$

Solution. We use Lagrangian

$$L = \frac{1}{2}(g_{11}z_x^2 + g_{22}z_y^2).$$

Compute the partial derivatives

$$\frac{\partial L}{\partial z} = 0, \quad \frac{\partial L}{\partial z_x} = 2g_{11}z_x, \quad \frac{\partial L}{\partial z_y} = 2g_{22}z_y$$

and make up Euler–Lagrange PDE

$$D_x(g_{11}z_x) + D_y(g_{22}z_y) = 0.$$

Replacing the function z, it remains an equation with partial derivatives with two unknowns g_{11} and g_{22}.

Problem 7.8.4: Find the string with minimum energy.

Solution. The energy of the function $z = z(x, t)$ is the functional

$$E = \frac{1}{2} \iint_{[0,1]^2} (z_x^2 + z_t^2) \, dx dt.$$

The vibrating spring is solution of PDE

$$z_{xx} - \frac{1}{c^2} z_{tt} = 0.$$

It follows the Lagrangian

$$L = \frac{1}{2}(z_x^2 + z_t^2) + p(x, t)\left(z_{xx} - \frac{1}{c^2} z_{tt}\right).$$

Since

$$\frac{\partial L}{\partial z} = 0, \; \frac{\partial L}{\partial z_x} = z_x, \; \frac{\partial L}{\partial z_t} = z_t, \; \frac{\partial L}{\partial z_{xx}} = p, \; \frac{\partial L}{\partial z_{tt}} = -\frac{p}{c^2}, \; \frac{\partial L}{\partial p} = z_{xx} - \frac{1}{c^2} z_{tt},$$

we find the Euler–Lagrange PDEs

$$-z_{xx} - z_{tt} + p_{xx} - \frac{1}{c^2} p_{tt} = 0, \; z_{xx} - \frac{1}{c^2} z_{tt} = 0.$$

We solve the second PDE and then from the first PDE we find the multiplier.

Problem 7.8.5: How many of the PDE solutions

$$\frac{\partial^2 z}{\partial x^2} - \frac{\partial^2 z}{\partial y^2} = 16(x^2 - y^2)$$

are extremals of the functional

$$I(z(\cdot)) = \frac{1}{2} \iint_{[0,1]^2} z_x z_y \, dx dy?$$

Solution. Since $L = \frac{1}{2} z_x z_y$, the Euler–Lagrange PDE is $z_{xy} = 0$. It follows the general solution $z = f(x) + g(y)$, where f and g are C^2 functions. Replacing in the restriction equation, we find

$$f''(x) - 16x^2 = g''(y) - y^2.$$

It follows $f''(x) - 16x^2 = a$ and $g''(y) - y^2 = a$. There are an infinity of functions f and g which satisfy these ODEs.

7.9 Multiple Integral Functionals With Nonholonomic Constraints

Problem 7.9.1: Let $\Omega_{t_0 t_1} \subset \mathbb{R}^2$ be a square. Find the surfaces

$$x^i = x^i(t), \ t = (t^1, t^2) \in \Omega_{t_0 t_1}, \ i = \overline{1,5},$$

in the distribution

$$D : dx^1(t) - x^2(t)dx^3(t) + x^4(t)dx^5(t) = 0, \ x(t_0) = x_0, \ x(t_1) = x_1$$

which have area 1.

Solution. The Pfaff equation is attached to a contact form. That's why the maximal integral manifolds have dimension two. The general solution of Pfaff equation is of the form $x^1 = f(x^3, x^5), x^2 = f_{x^3}, x^4 = f_{x^5}$, where f is an arbitrary C^2 function. Let $x^i = x^i(t^1, t^2), i = 1, ..., 5$, be a parametrization of the surface, solution of Pfaff equation. Let $h_{\alpha\beta} = \delta_{ij} x_\alpha^i x_\beta^j$, $\alpha, \beta = 1, 2$, the metric induced by the metric δ_{ij} in $\mathbb{R}^5$. The area of this surface is

$$\sigma = \iint_{\Omega_{t_0 t_1}} \sqrt{\det(h_{\alpha\beta})} \ dt^1 dt^2.$$

It is required $\sigma = 1$.

Problem 7.9.2: There are curvilinear triangles Δ with boundary (the three sides, curves) in the distribution

$$(x + y)dx + (z - x)dy - zdz = 0$$

and with vertices $(1, 0, 0), (0, 1, 0), (0, 0, 1)$, which optimize the functional

$$I(\Delta) = \int_{\Delta_1} dx dy,$$

where Δ_1 is a curvilinear convex triangle, projection on the plane xOy of Δ?

Solution. Dividing by dt, we transform the Pfaff equation in the ODE

$$(x + y)\dot{x} + (z - x)\dot{y} - z\dot{z} = 0,$$

with three unknown functions $x(t), y(t), z(t), t \in I \subset \mathbb{R}$ (curve of class C^1). Suppose that the three sides of the triangle Δ are the curves

$$C_1 : (x_1(t), y_1(t), z_1(t)); C_2 : (x_2(t), y_2(t), z_2(t)); C_3 : (x_3(t), y_3(t), z_3(t)),$$

which verify the ODE and each joins two given points.

In projection, the convex triangle Δ_1 has the vertices $(0, 0), (1, 0), (0, 1)$ and sides

$$c_1 : (x_1(t), y_1(t)); c_2 : (x_2(t), y_2(t)); c_3 : (x_3(t), y_3(t)).$$

The smallest area of the triangle Δ_1 is made when c_1, c_2, c_3 are straight lines.

Problem 7.9.3: Compute the first variation of the functional

$$I(y(\cdot)) = \iint_{[0,1]^2} y(x)y(t) \sin(xt)\, dxdt,$$

where $tdy + (t - y)dt = 0$.

Solution. We consider the variation $\bar{y}(x) = y(x) + \varepsilon h(x)$ which satisfies the restriction, i.e.,

$$t(dy + \varepsilon dh) + (t - y - \varepsilon h)dt = 0.$$

Derive with respect to ε and set $\varepsilon = 0$. We find $tdh - hdt = 0$, i.e., $h = c_1 t$. So $\bar{y}(x) = y(x) + \varepsilon c_1 x$. We deduce

$$I(\varepsilon) = \iint_{[0,1]^2} \sin(xt)(y(x) + \varepsilon c_1 x)(y(t) + \varepsilon c_1 t)\, dxdt.$$

The first variation is

$$I'(0) = 2c_1 \iint_{[0,1]^2} \sin(xt)(y(x)t + y(t)x)\, dxdt.$$

Problem 7.9.4: Let Ω be a cube in $\mathbb{R}^3$. Find the extremals of the functional

$$I(u(\cdot)) = \int_{\Omega} (u + u_x + u_y + u_z)\, dxdydz$$

constrained by Pfaff equation $dx + dy + xydz = 0$.
Hint We can use the Lagrangian 3-form

$$L = (u + u_x + u_y + u_z)\, dx \wedge dy \wedge dz + \omega \wedge (dx + dy + xydz),$$

where the multiplier is a 2-form

$$\omega = p(x,y,z)\, dy \wedge dz + q(x,y,z)\, dz \wedge dx + r(x,y,z)\, dx \wedge dy.$$

Problem 7.9.5: There are three solutions to the Pfaff equation $dz = \frac{1}{2} y^2 dx$ what fixes a convex curvilinear cube of minimum volume?
Hint. The general solution of the Pfaff equation can be written in the form $x = \int f(t)dt, y = 2t, z = \int 2t^2 f(t)dt$, where f is an arbitrary C^1 function.

7.10 Multiple Integral Functionals With Isoperimetric Constraints

Problem 7.10.1: Let $\Omega = [0, 1]^2 \subset \mathbb{R}^2$. Find the extremals of the functional

$$I(u(\cdot)) = \frac{1}{2} \int_{\Omega} (u_x^2 + u_y^2)\, dxdy,$$

subject to the restriction

$$\int_\Omega u_x u_y \, dxdy = 1$$

and to boundary conditions $u(0,0) = 0$, $u(1,1) = 1$.

Problem 7.10.2: Let $\Omega = [0,1]^2 \subset \mathbb{R}^2$ and Γ be a curve in Ω joining the points $(0,0)$ and $(1,1)$. Find the extremals of the functional

$$I(u(\cdot)) = \frac{1}{2} \int_\Omega (u_x^2 + u_y^2) \, dxdy,$$

with the constraint

$$\int_\Gamma u_{xy} \, d(xy) = 0,$$

path-independent curvilinear integral, and with boundary conditions $u(0,0) = 0$, $u(1,1) = 1$.

Solution. The path independence of the curvilinear integral is ensured if and only if there is a function $v(xy)$ of class C^1 such that $u_{xy}(x,y) = v(xy)$.

We introduce the 2-form $\omega = dx \wedge dy$, the 1-forms $\omega_1 = dy$, $\omega_2 = -dx$ and the multiplier 1-form $p = p^1 \omega_1 + p^2 \omega_2$. We build the Lagrangian 2-form

$$L = \frac{1}{2}(u_x^2 + u_y^2) \, dx \wedge dy + (p^1 \omega_1 + p^2 \omega_2) \wedge (yu_{xy} dx + xu_{xy} dy)$$

$$= \frac{1}{2}(u_x^2 + u_y^2) \, dx \wedge dy + (p^1 dy - p^2 dx) \wedge (yu_{xy} dx + xu_{xy} dy)$$

$$= \left(\frac{1}{2}(u_x^2 + u_y^2) - (p^1 y + p^2 x)u_{xy} \right) dx \wedge dy.$$

The Lagrangian coefficient

$$\mathcal{L} = \frac{1}{2} \, (u_x^2 + u_y^2) - (p^1 y + p^2 x) \, u_{xy}$$

produces the Euler–Lagrange equations

$$\frac{\partial \mathcal{L}}{\partial u} - D_x \frac{\partial \mathcal{L}}{\partial u_x} - D_y \frac{\partial \mathcal{L}}{\partial u_y} + D_{xy} \frac{\partial \mathcal{L}}{\partial u_{xy}} = 0, \quad u_{xy} = 0$$

or

$$u_{xx} + u_{yy} + D_{xy}(p^1 y + p^2 x) = 0, \quad u_{xy} = 0.$$

The general solution of the last PDE is $u = f(x) + g(y)$, with f and g functions of class C^2. Replacing in the first equation, it follows the multipliers.

Problem 7.10.3: Let $\Omega = [0,1]^2 \subset \mathbb{R}^2$. Find the extremals of the functional

$$I(u(\cdot)) = \frac{1}{2} \int_\Omega (u_t^2 - u_x^2)\, dt dx,$$

with the constraint

$$\int_\Omega (u_t + u_x)\, dt dx = 1$$

and with boundary conditions $u(0,0) = 0$, $u(1,1) = 1$.

Problem 7.10.4: Let $\Omega = [0,1]^2 \subset \mathbb{R}^2$ and Γ be a curve in Ω which joins the points $(0,0)$ and $(1,1)$. Find the extremals of the functional

$$I(u(\cdot)) = \frac{1}{2} \int_\Omega (u_t^2 - u_x^2)\, dt dx,$$

subject to the constraint

$$\int_\Gamma (u_t + u_x)\, d(t+x) = 1,$$

path-independent curvilinear integral, and with the boundary conditions $u(0,0) = 0$, $u(1,1) = 1$.

Problem 7.10.5: Let D be a bounded plane domain and $v : D \to \mathbb{R}$ be a given continuous function. Determine the extremals of the functional

$$I(u(\cdot)) = \frac{1}{2} \int_D (u_x^2 + u_y^2)\, dx dy,$$

subject to constraints

$$\frac{1}{2} \int_D u^2\, dx dy = 1, \quad \int_D uv\, dx dy = 0$$

and to boundary conditions $u(0,0) = 0$, $u(1,1) = 1$.

7.11 Curvilinear Integral Functionals With PDE Constraints

Problem 7.11.1: Let $\Omega = [0,1]^2 \subset \mathbb{R}^2$ and Γ a curve in Ω which joins the points $(0,0)$ and $(1,1)$. Determine the extremals of the functional

$$I(u(\cdot)) = \frac{1}{2} \int_\Gamma u_{t^1}^2(t^1, t^2)\, dt^1 - u_{t^2}^2(t^1, t^2)\, dt^2,$$

under the following simultaneous conditions: (1) path-independent curvilinear integral, (2) constraint $u_{t^1} + 2u_{t^2} + u = 0$ and (3) two boundary conditions $u(0,0) = 1$, $u(1,1) = 1$.

Solution. The characteristic system associated to the restriction equation is

$$\frac{dt^1}{1} = \frac{dt^2}{2} = -\frac{du}{u}.$$

It follows the first integrals $2t^1 - t^2 = c_1$, $u = c_2 e^{-t^1}$ and the general solution $u = e^{-t^1} \phi(2t^1 - t^2)$, where ϕ is a function of class C^1.

On the other hand, the Lagrangian 1-form (L_1, L_2) leads to Euler–Lagrange equations

$$\frac{\partial L_1}{\partial u} - D_\gamma \frac{\partial L_1}{\partial u_\gamma} = 0, \quad \frac{\partial L_2}{\partial u} - D_\gamma \frac{\partial L_2}{\partial u_\gamma} = 0.$$

For Lagrangian 1-form

$$L_1 = \frac{1}{2} u_{t^1}^2, \ L_2 = -\frac{1}{2} u_{t^2}^2$$

we find the equations

$$u_{t^1 t^1} + u_{t^1 t^2} = 0, \ u_{t^1 t^2} + u_{t^2 t^2} = 0.$$

These imply

$$\phi - 3\phi' + 2\phi'' = 0, \ \phi' - \phi'' = 0.$$

It follows $\phi'' = \phi = \phi'$. We find $\phi = ke^\xi$, $\xi = 2t^1 - t^2$. Finally $u = e^{t^1 - t^2}$.

Problem 7.11.2: Let $\Omega = [0, 1]^2 \subset \mathbb{R}^2$ and Γ be a curve in Ω joining the points $(0, 0)$ and $(1, 1)$. Determine the extremals of the functional

$$I(u(\cdot)) = \frac{1}{2} \int_\Gamma u_{t^1 t^2}(dt^1 + dt^2),$$

path-independent curvilinear integral, with constraint $2u_{t^1} + 3u_{t^2} - u = 0$ and boundary conditions $u(0, 0) = 0$, $u(1, 1) = 1$.

Problem 7.11.3: Let $\Omega = [0, 1]^2 \subset \mathbb{R}^2$ and Γ be a curve in Ω joining the points $(0, 0)$ and $(1, 1)$. Determine the extremals of the functional

$$I(u(\cdot)) = \int_\Gamma u_{t^1}^2(t^1, t^2)\, dt^1 + u_{t^2}^2(t^1, t^2)\, dt^2,$$

path-independent curvilinear integral, constrained by $u_{t^1} + 2u_{t^2} + u = 0$ and with boundary conditions $u(0, 0) = 0$, $u(1, 1) = 1$.

Problem 7.11.4: Assume $\Omega = [0, 1]^2 \subset \mathbb{R}^2$ and let Γ be a curve in Ω joining the points $(0, 0)$ and $(1, 1)$. Determine the extremals of the functional

$$I(u(\cdot)) = \frac{1}{2} \int_\Gamma u_{t^1 t^2}(dt^1 + dt^2),$$

path-independent curvilinear integral, constrained by $2u_{t^1} + 3u_{t^2} - u = 0$ and with boundary conditions $u(0, 0) = 0$, $u(1, 1) = 1$.

Problem 7.11.5: Assume $\Omega = [0, 1]^2 \subset \mathbb{R}^2$ and let Γ be a curve in Ω joining the points $(0, 0)$ and $(1, 1)$. Determine the extremals of the functional

$$I(u(\cdot)) = \frac{1}{2} \int_\Gamma u_{t^1 t^2} \, d(t^1 t^2),$$

path-independent curvilinear integral, constrained by $t^1 u_{t^1} + t^2 u_{t^2} - u = 0$ and with boundary conditions $u(0, 0) = 0$, $u(1, 1) = 1$.

7.12 Curvilinear Integral Functionals With Nonholonomic Constraints

Problem 7.12.1: Find the surfaces $x^i = x^i(t)$, $t = (t^1, t^2) \in \Omega_{t_0 t_1} \subset \mathbb{R}^2$, $i = \overline{1, 5}$, in the distribution

$$D : dx^1(t) - x^2(t)dx^3(t) + x^4(t)dx^5(t) = 0, \quad x(t_0) = x_0, \; x(t_1) = x_1,$$

whose boundaries are closed curves of length 1.

Problem 7.12.2: Assume $\Omega = [0, 1]^5 \subset \mathbb{R}^5$ and let Γ be a curve in Ω joining the points $0 = (0, 0, 0, 0, 0)$ and $1 = (1, 1, 1, 1, 1)$. Determine the extremals of the functional

$$I(x(\cdot)) = \int_\Gamma dx^1 + x^2 dx^3,$$

path-independent curvilinear integral, with constraints

$$D : dx^1 - x^2 dx^3 + x^4 dx^5 = 0, \quad x(0) = x_0, \; x(1) = x_1.$$

Hint. The general solution of Pfaff equation is

$$x^1 = f(x^3, x^5), \quad x^2 = f_{x^3}, \quad x^4 = -f_{x^5},$$

where f is an arbitrary C^2 function.

Problem 7.12.3: Assume $\Omega = [0, 1]^3 \subset \mathbb{R}^3$ and let Γ be a curve in Ω joining the points $0 = (0, 0, 0)$ and $1 = (1, 1, 1)$. Determine the extremals of the functional

$$I(x(\cdot), y(\cdot), z(\cdot)) = \int_\Gamma f(x, y, z) \, d(xyz),$$

path-independent curvilinear integral, subject to the nonholonomic constraint $dz = ydx$.

Solution. The path independence conditions (complete integrability)

$$\frac{\partial}{\partial y}(xyf) - \frac{\partial}{\partial z}(xzf) = 0, \quad \frac{\partial}{\partial z}(yzf) - \frac{\partial}{\partial x}(yxf) = 0, \quad \frac{\partial}{\partial x}(yzf) - \frac{\partial}{\partial y}(yxf) = 0$$

imply the existence of a function g of class C^1 such that $f(x, y, z) = g(xyz)$.

We introduce the curve $x = x(t), y = y(t), z = z(t), t \in [a, b]$ of class C^1, which joins the given points. Then the functional becomes a simple integral

$$I = \int_a^b g(x(t)y(t)z(t))(y(t)z(t)\dot{x}(t) + z(t)x(t)\dot{y}(t) + x(t)y(t)\dot{z}(t)) \, dt,$$

and the restriction writes $\dot{z}(t) = y(t)\dot{x}(t)$. Eliminating $\dot{z}$, we find the functional

$$J = \int_a^b g(x(t)y(t)z(t))(y(t)z(t)\dot{x}(t) + z(t)x(t)\dot{y}(t) + x(t)y^2(t)\dot{x}(t)) \, dt,$$

under constrained $\dot{z}(t) = y(t)\dot{x}(t)$. The Lagrangian

$$L = g(x(t)y(t)z(t))(y(t)z(t)\dot{x}(t) + z(t)x(t)\dot{y}(t) + x(t)y^2(t)\dot{x}(t))$$

is replaced by a new Lagrangian

$$\mathcal{L} = g(x(t)y(t)z(t))(y(t)z(t)\dot{x}(t) + z(t)x(t)\dot{y}(t) + x(t)y^2(t)\dot{x}(t))$$
$$+ p(t)(\dot{z}(t) - y(t)\dot{x}(t)).$$

The Lagrangian L has the partial derivatives

$$\frac{\partial L}{\partial x} = g'yz(yz\dot{x} + zx\dot{y} + xy^2\dot{x}) + g(z\dot{y} + y^2\dot{x}), \quad \frac{\partial L}{\partial \dot{x}} = g(yz + xy^2);$$

$$\frac{\partial L}{\partial y} = g'xz(yz\dot{x} + zx\dot{y} + xy^2\dot{x}) + g(z\dot{x} + 2xy\dot{x}), \quad \frac{\partial L}{\partial \dot{y}} = gxz;$$

$$\frac{\partial L}{\partial z} = g'xy(yz\dot{x} + zx\dot{y} + xy^2\dot{x}) + g(y\dot{x} + xy), \quad \frac{\partial L}{\partial \dot{z}} = 0.$$

The extremals are solutions of Euler–Lagrange equations written with $\mathcal{L}$.

Problem 7.12.4: Find the curves Γ in the distribution $dz = \frac{1}{2} y^2 dx$ which optimize the functional $I(\Gamma) = \int_\Gamma f(x, y, z) \, d(xyz)$.

Hint. The general solution of the Pfaff equation can be written in the form $x = \int f(t)dt, y = 2t, z = \int 2t^2 f(t)dt$, where f is an arbitrary function.

Problem 7.12.5: Assume $\Omega = [0, 1]^5 \subset \mathbb{R}^5$ and let Γ be a curve in Ω joi- ning the points $0 = (0, 0, 0, 0, 0)$ and $1 = (1, 1, 1, 1, 1)$. Determine the extremals of the functional

$$I(x(\cdot)) = \int_\Gamma f(x^1, x^2, x^3, x^4, x^5) \, d(x^1x^2x^3x^4x^5),$$

path-independent curvilinear integral, with constraints

$$D : dx^1 - x^2dx^3 + x^4dx^5 = 0, \quad x(0) = x_0, \quad x(1) = x_1.$$

Remark 7.5: The following properties are equivalent: (i) $\int_\Gamma P(x, y, z)dx + Q(x, y, z)dy + R(x, y, z)dz$ does not depend on the path of integration; (ii) There is a function $u = u(x, y, z)$ so that $du = P(x, y, z)dx + Q(x, y, z)dy + R(x, y, z)dz$ and $\int_A^B P(x, y, z)dx +$

$Q(x, y, z)dy + R(x, y, z)dz = u(B) - u(A)$; (iii) $\frac{\partial P}{\partial y} - \frac{\partial Q}{\partial x} = 0$, $\frac{\partial P}{\partial z} - \frac{\partial R}{\partial x} = 0$, $\frac{\partial Q}{\partial z} - \frac{\partial R}{\partial y} = 0$; (iv)
$\int_\Gamma P(x, y, z)dx + Q(x, y, z)dy + R(x, y, z)dz = 0$ if the curve Γ is closed.

7.13 Curvilinear Integral Functionals with Isoperimetric Constraints

Problem 7.13.1: Assume $\Omega = [0, 1]^2 \subset \mathbb{R}^2$ and let Γ be a curve in Ω joining the points $(0, 0)$ and $(1, 1)$. Determine the extremals of the functional

$$I(u(\cdot)) = \int_\Gamma \frac{1}{2}(u_{t^1}^2 + u_{t^2}^2)\, dt^1 + u_{t^1} u_{t^2}\, dt^2,$$

path-independent curvilinear integral, with constraint

$$\int_\Gamma u_{t^1}^2\, dt^1 + u_{t^2}^2\, dt^2 = 1,$$

path-independent curvilinear integral, and with boundary conditions $u(0, 0) = 0$, $u(1, 1) = 1$.

Problem 7.13.2: Assume $\Omega = [0, 1]^2 \subset \mathbb{R}^2$ and let Γ be a curve in Ω joining the points $(0, 0)$ and $(1, 1)$. Find the extremals of the functional

$$I(u(\cdot)) = \int_\Gamma \frac{1}{2}(u_{t^1}^2 + u_{t^2}^2)\, dt^1 + u_{t^1} u_{t^2}\, dt^2,$$

path-independent curvilinear integral, with constraint

$$\int_\Omega u_{t^1} u_{t^2}\, dt^1 dt^2 = 1$$

and with the boundary conditions $u(0, 0) = 0$, $u(1, 1) = 1$.

Problem 7.13.3: Assume $\Omega = [0, 1]^2 \subset \mathbb{R}^2$ and let Γ be a curve in Ω joining the points $(0, 0)$ and $(1, 1)$. Find the extremals of the functional

$$I(u(\cdot)) = \int_\Gamma u_{t^1}^2\, dt^1 + u_{t^2}^2\, dt^2,$$

path-independent curvilinear integral, subject to the constraint

$$\int_\Gamma u\, dt^1 + dt^2 = 1,$$

path-independent curvilinear integral, and to boundary conditions $u(0, 0) = 0$, $u(1, 1) = 1$.

Problem 7.13.4: Assume $\Omega = [0,1]^2 \subset \mathbb{R}^2$ and let Γ be a curve in Ω joining the points $(0,0)$ and $(1,1)$. Find the extremals of the functional

$$I(u(\cdot)) = \int_\Gamma u_{t^1}^2 \, dt^1 - u_{t^2}^2 \, dt^2,$$

path-independent curvilinear integral, subject to the restriction

$$\int_\Omega (u - u_{t^1} - u_{t^2}) \, dt^1 dt^2 = 1$$

and to boundary conditions $u(0,0) = 0$, $u(1,1) = 1$.

Solution. Here we need to use a vector multiplier

$$p = p^1 \frac{\partial}{\partial t^1} + p^2 \frac{\partial}{\partial t^2}$$

and the interior product to create the Lagrangian 1-form

$$L = u_{t^1}^2 \, dt^1 - u_{t^2}^2 \, dt^2 + \left(p^1 \frac{\partial}{\partial t^1} + p^2 \frac{\partial}{\partial t^2} \right) \rfloor ((u - u_{t^1} - u_{t^2}) \, dt^1 \wedge dt^2)$$

$$= (u_{t^1}^2 - p^2(u - u_{t^1} - u_{t^2})) \, dt^1 + (-u_{t^2}^2 + p^1(u - u_{t^1} - u_{t^2})) \, dt^2.$$

The Lagrangian 1-form

$$L_1 = u_{t^1}^2 - p^2(u - u_{t^1} - u_{t^2}), \quad L_2 = -u_{t^2}^2 + p^1(u - u_{t^1} - u_{t^2})$$

produces two Euler–Lagrange equations

$$\frac{\partial L_\alpha}{\partial u} - D_\gamma \frac{\partial L_\alpha}{\partial u_\gamma} = 0, \ \alpha = 1, 2,$$

with sum over $\gamma = 1, 2$.

Variant We apply the principle of reciprocity (we optimize the functional from restriction with constraint given by initial functional). We introduce the 2-form $\omega = dt^1 \wedge dt^2$, 1-forms

$$\omega_\lambda = \frac{\partial}{\partial t^\lambda} \rfloor \omega \ : \ \omega_1 = dt^2, \ \omega_2 = -dt^1$$

and the multiplier 1-form $p = p^\lambda \omega_\lambda$. We build the Lagrangian 2-form

$$L = (u - u_{t^1} - u_{t^2}) dt^1 \wedge dt^2 + p^\lambda \omega_\lambda \wedge (u_{t^1}^2 \, dt^1 - u_{t^2}^2 \, dt^2)$$

$$= (u - u_{t^1} - u_{t^2} - p^1 u_{t^1}^2 + p^2 u_{t^2}^2) \, dt^1 \wedge dt^2.$$

The Lagrangian

$$\mathcal{L} = u - u_{t^1} - u_{t^2} - p^1 u_{t^1}^2 + p^2 u_{t^2}^2$$

produces Euler–Lagrange equations

$$\frac{\partial \mathcal{L}}{\partial u} - D_\gamma \frac{\partial \mathcal{L}}{\partial u_\gamma} = 0, \quad \frac{\partial p^1}{\partial t^1} + \frac{\partial p^2}{\partial t^2} = 0$$

or

$$1 + 2\frac{\partial}{\partial t^1}(p^1 u_{t^1}) - 2\frac{\partial}{\partial t^2}(p^2 u_{t^2}) = 0, \quad \frac{\partial p^1}{\partial t^1} + \frac{\partial p^2}{\partial t^2} = 0.$$

Problem 7.13.5: Let $D : x^2 + y^2 \le 1$ be a plane domain, Γ be a curve in D joining the points $(0,0)$ and $(\frac{1}{2}, \frac{1}{2})$ and $v : D \to \mathbb{R}$ be a given continuous function. Determine the extremals of the functional

$$I(u(\cdot)) = \int_\Gamma u_x^2 \, dx - u_y^2 \, dy,$$

with restrictions

$$\int_D (u^2 + x^2 + y^2) \, dxdy = 1, \quad \int_D uv \, dxdy = 0$$

and with boundary conditions $u(0,0) = 0$, $u\left(\frac{1}{2}, \frac{1}{2}\right) = 1$.

7.14   Maple Application Topics

Maple is the world leader in finding exact solutions to ordinary and partial differential equations.

Problem 7.14.1:   Solving an ODE:
> ode := $diff(y(x), x, x) = 2 * y(x) + 1$;
> dsolve(ode);
> ics := $y(0) = 1, (D(y))(0) = 0$;
> dsolve(ics, ode);

Problem 7.14.2:   Solving a PDE:
> PDE := $diff(u(x,t), t) = -diff(u(x,t), x)$;
subject to the initial-boundary conditions:
> IBC := $u(x,0) = sin(2 * Pi * x), u(0,t) = -sin(2 * Pi * t)$;
Then the numerical solution is given by
> pds := pdsolve(PDE,IBC,numeric,time=t,range=0..1);

Problem 7.14.3:   > with(PDEtools, casesplit, declare);
> with(DEtools, gensys);
> declare(F(r, s), H(r), G(s));
> sys2 := $[-(diff(F(r,s),r,r)) + diff(F(r,s),s,s)$
$+diff(H(r),r) + diff(G(s),s) + s = 0,$
$diff(F(r,s),r,r) + 2 * (diff(F(r,s),s,r)) + diff(F(r,s),s,s)$
$-(diff(H(r),r)) + diff(G(s),s) - r = 0]$

> sol := pdsolve(sys2);

$$F = F1(s) + F2(s) + F3(s - r) - (1/12) * r^2 * (r - 3 * s),$$
$$G(s) = -(diff(F1(s), s)) - (1/4) * s^2 + C2,$$
$$H(r) = diff(F2(r), r) - (1/4) * r^2 + C1$$

and depends on three arbitrary functions,

$$F1(s), F2(r), F3(s - r)$$

> pdetest(sol, sys2)
$$[0, 0]$$

Note that the numerical solution of a pde returns a module, not a proc like the numerical solution of an ode.

The VariationalCalculus package contains commands that perform calculus of variations computations. This package performs various computations for characterizing the extremals of functionals. Each command in the VariationalCalculus package can be accessed by using either the long form or the short form of the command name in the command calling sequence.

Problem 7.14.4: **(The mirage)** Find the extremal curve to

$$I(y(\cdot)) = \int_{x_0}^{x_1} (1 + ky(x)) n_0 \sqrt{1 + y'(x)^2} \, dx,$$

where $(1 + ky(x)) n_0$ is the index of refraction. Try to get the results through Maple packages.

Problem 7.14.5: **(Eigenfrequency)** Find the extremal curve $y(x)$ of the functional

$$I(y(\cdot)) = \frac{1}{2} \int_0^1 y'(x)^2 \, dx, \quad y(0) = 0, y(1) = 0,$$

subject to the constraint

$$I(y(\cdot)) = \frac{1}{2} \int_0^1 y(x)^2 \, dx = 1.$$

Try to get the results through Maple packages.

Answer: the extremal curves are $y_k = 2 \sin k\pi x$, with the corresponding Lagrange multipliers $\lambda_k = k^2 \pi^2$.

Problem 7.14.6: (**Fluid mechanics**) Find the extremals of the energy functional

$$I(u(\cdot)) = \frac{1}{2} \int_\Omega (u_t^2 + u_x^2) \, dt dx$$

constrained by Bateman–Burgers PDE

$$u_t + u u_x = \nu u_{xx}.$$

Try to get the results through Maple packages.

Problem 7.14.7: Find the extremals of the hyperbolic energy functional

$$J(u(\cdot)) = \frac{1}{2} \int_\Omega (u_t^2 - u_x^2) \, dt dx$$

constrained by Chafee-Infante PDE

$$u_t - u_{xx} + \lambda(u^3 - u) = 0.$$

Try to get the results through Maple packages.

Problem 7.14.8: Determine the extremum of the functional

$$I(z(\cdot)) = \frac{1}{2} \iint_{[0,1]^2} (z_x^2 + z_y^2) \, dx dy,$$

constrained by isoperimetric condition

$$J = \iint_{[0,1]^2} (z_x z_y - z) \, dx dy = 1.$$

Try to get the results through Maple packages.

Problem 7.14.9: Find the extremals of the hyperbolic energy functional

$$I(u(\cdot)) = \frac{1}{2} \int_{[0,1]^2} (u_t^2 - u_x^2) \, dt dx$$

subject to

$$J(u(\cdot)) = \iint_{[0,1]^2} (u_t - u_{xx} + \lambda(u^3 - u)) \, dt dx = 1.$$

Try to get the results through Maple packages.

Problem 7.14.10: Minimize the functional

$$I(u) = \int_\Omega F(x, u(x), \nabla u(x)) \, dx^1 ... dx^n$$

subject to the isoperimetric constraint

$$J(u) = \int_\Omega G(x, u(x), \nabla u(x)) \, dx^1 ... dx^n = 0.$$

Hint. The minimizer u (if exists) satisfies the Euler–Lagrange equation

$$\left(\left\langle \nabla, \frac{\partial}{\partial \nabla u} \right\rangle - \frac{\partial}{\partial u} \right)(F + \lambda G) = 0$$

and the natural boundary condition

$$\left\langle n, \frac{\partial}{\partial \nabla u} \right\rangle (F + \lambda G) = 0, \quad \text{on } \partial\Omega.$$

Problem 7.14.11: (**Dirichlet minimization problem**) Let $\Omega \subset \mathbb{R}^2$ be a square homo-geneous domain. Find $\min_u \int_\Omega |\nabla u|^2 \, dxdy$ subject to the constraint (curvilinear integral of first type)

$$\int_{\partial\Omega_1} u \, ds - \int_{\partial\Omega_2} u \, ds = V,$$

where $\partial\Omega_1$ is the left part and $\partial\Omega_2$ is the right part of the boundary $\partial\Omega$.

Problem 7.14.12: Let $\Omega \subset \mathbb{R}^2$ be a square domain. Minimize the surface area integral

$$I(z) = \int_\Omega \sqrt{1 + z_x^2 + z_y^2} \, dxdy$$

constrained by

$$\int_{\partial\Omega} z \, ds = 1.$$

Problem 7.14.13: Let Ω be a regular domain in $\mathbb{R}^n$. Establish the existence of minimizers in $W^{1,2}(\Omega)$ of the functional

$$\int_\Omega (\|\nabla u\|^2 + |u|^2) \, dx^1 ... dx^n + \int_{\partial\Omega} g(x) u \, d\sigma,$$

where $g(x) \in L^2(\partial\Omega)$.

Problem 7.14.14: The ODEs of motion of a nonholonomic constrained system are reduced to

$$\ddot{q}^1 = \frac{-1}{2t} \dot{q}^1, \quad \ddot{q}^2 = \frac{-1}{2t} \dot{q}^2, \quad \dot{q}^3(t) = \sqrt{\frac{1}{t} - \dot{q}^1(t)^2 - \dot{q}^2(t)^2} \,.$$

Show that the general solution of this ODEs system is

$$q^1(t) = a_1^1 \sqrt{t} + a_2^1, \quad q^2(t) = a_1^2 \sqrt{t} + a_2^2, \quad q^3(t) = a_1^3 \sqrt{t} + a_2^3,$$

where a_j^i are constants connected by the relation $a_1^3 = \sqrt{4 - (a_1^1)^2 + (a_1^2)^2}\,.$

Problem 7.14.15: The unicycle configuration space has three dimensions: x, y, θ. We can write

$$\frac{d}{dt} \begin{pmatrix} x \\ y \\ \theta \end{pmatrix} = \lambda \begin{pmatrix} 0 \\ 0 \\ 1 \end{pmatrix} + \mu \begin{pmatrix} \cos\theta \\ \sin\theta \\ 0 \end{pmatrix}.$$

Here λ tells us how the unicyclist is steering (changing θ), while μ gives the forward motion where "forward" is the direction that θ is pointing. Find the geometric dynamics generated by this ODEs system and Euclidean metric.

Problem 7.14.16: A particle is constrained to move on a sphere in three-dimensional space whose radius changes with time t: $x\,dx + y\,dy + z\,dz - c^2 dt = 0$. If the kinetic energy is given by $T = \frac{1}{2}(\dot{x}^2 + \dot{y}^2 + \dot{z}^2)$, find Euler–Lagrange equations of movement.

Problem 7.14.17: Let us consider the double integral functional

$$I(u) = \int_\Omega \frac{1}{1 + u_{x^1}^2 + u_{x^2}^2} \, dx^1 \wedge dx^2,$$

where Ω is the support of the function u. Find the Hamiltonian H, write the discrete version of H and realize a numerical study of the motion.

Problem 7.14.18: Let us consider a domain $\Omega \subset \mathbb{R}^n$, a point $x \in \Omega$, and the functional $I(w) = \int_\Omega L(x, w, Dw) \, dx^1 ... dx^n$. Find the Lagrangian L such that $-\Delta w + \langle D\varphi, DW \rangle = f$ is the Euler–Lagrange PDE corresponding to the functional $I(w)$.

Solution. Δw is the divergence of Dw. On the other hand

$$\operatorname{div}(e^{-\varphi} Dw) = e^{-\varphi}(\langle -D\varphi, Dw \rangle + \Delta w).$$

It remains to create $e^{-\varphi} f$ in the equation, which we leave for reader. Finally, $L(x, w, p) = e^{-\varphi(x)}(\frac{1}{2}|p|^2 - f(x)w)$.

Bibliography

[1] C. M. Arizmendi, J. Delgado, H. N. Núñez-Yépez, A. L. Salas-Brito, *Lagrangian description of the variational equations*, arXiv:math-ph/0403041v1, 21 Mar 2004.

[2] P. Ballard, *The dynamics of discrete mechanical systems with perfect unilateral constraints*, Archive for Rational Mechanics and Analysis, 154 (2000), 199–274.

[3] A. Bossavit, *Differential forms and the computation of fields and forces in electromagnetism*, European Journal of Mechanics – B/Fluids, 10, 5 (1991), 474–488.

[4] C. J. Budd, M. D. Piggott, *Geometric Integration and Its Applications*, Handbook of Numerical Analysis, Vol. XI, North-Holland Publishing Co., Amsterdam, 2003, pp. 35–139.

[5] O. Calin, D.-C. Chang, S. S. T. Yau, *Nonholonomic systems and sub-Riemannian geometry*, Communications in Information Systems, 10, 4 (2010), 293–316.

[6] K. W. Cassel, *Variational Methods with Applications in Science and Engineering*, Cambridge University Press, Cambridge, 2013.

[7] F. H. Clarke, *Calculus of Variations and Optimal Control Lecture Notes*, UBC, Vancouver, 1979.

[8] R. Courant, D. Hilbert, *Methods of Mathematical Physics (v. I)*, Interscience Publishers, Inc., New York, 1966.

[9] R. Courant, K. Friedrichs, H. Lewy, *On the partial difference equations of mathematical physics*, IBM Journal of Research and Development, 11, 2 (1967 [1928]), 215–234.

[10] C. Cronström, T. Raita, *On non-holonomic systems and variational principles*, arXiv:0810.3611v1 [physics.class-ph], 2008.

[11] S. Dutta, *Lecture Notes on Dynamic Meteorology*, India Meteorological Department Meteorological Training Institute, Pashan, Pune-8, 2021, https://imdpune.gov.in, training notes, FTC1.

[12] Eduardo Souza de Cursi, *Variational Methods for Engineers with Matlab*, John Wiley & Sons, Inc., Chichester, 2015.

[13] L. C. Evans, D. Gomes, *Effective Hamiltonians and averaging for Hamiltonian dynamics I*, Archive for Rational Mechanics and Analysis, 157, 1 (2001), 1–33.

[14] L. C. Evans, *Partial Differential Equations*, American Mathematical Society, Providence, RI, 1998.

Variational Calculus with Engineering Applications, First Edition. Constantin Udriste and Ionel Tevy.
© 2023 John Wiley & Sons Ltd. Published 2023 by John Wiley & Sons Ltd.

[15] C. Fox, *An Introduction to the Calculus of Variations*, Dover Publications, Inc., New York, 1963.

[16] I. M. Gelfand, S. V. Fomin, *Calculus of Variations*, Dover, New York, 2000.

[17] C. Ghiu, C. Udriste, L. L. Petrescu, *Families of solutions of multitemporal nonlinear Schrödinger PDE*, Mathematics, 9, 16 (2021), 1995.

[18] F. Giannessi, A. Maugeri (Editors), *Variational Analysis and Applications*, Springer, New York, 2005.

[19] M. Giaquinta, *Multiple integrals in the calculus of variations and nonlinear elliptic systems*, Annals of Mathematics Studies 105. Princeton University Press, Princeton, NJ, 1983.

[20] D. Glickenstein, *Introduction to Flows on Riemannian Metrics*, Math. 538, Springer, New York, 2009.

[21] H. Goldstine, *A History of the Calculus of Variations from the 17th Through the 19th Century*, Springer Science & Business Media, New York, 2012.

[22] D. A. Gomes, *Calculus of Variations and Partial Differential Equations*, https://www.math.tecnico.ulisboa.pt/~dgomes/notas_calvar.pdf, 2022.

[23] Y.-X. Guo, C. Liu, S.-X. Liu, *Generalized Birkhoffian realization of noholonomic systems*, Communications in Mathematics, 18 (2010), 21–35.

[24] H. Y. Guo, Y. Q. Li, K. Wu, S. K. Wang, *Difference discrete variational principle, Euler–Lagrange cohomology and symplectic, multisymplectic structures*, arXiv:math-ph/0106001v1, 2001, DOI: 10.1088/0253-6102/37/2/129.

[25] E. Hairer, C. Lubich, G. Wanner, *Geometric Numerical Integration: Structure-Preserving Algorithms for Ordinary Differential Equations*, Springer-Verlag, New York, 2002.

[26] Q. Huang, *Solving an Open Sensor Exposure Problem using Variational Calculus*, Report Number: WUCSE-2003-1, 2003. All Computer Science and Engineering Research.

[27] H. Härtel, M. Lüdke, *3D-simulations of interacting particles*, Computing in Science and Engineering, 4 (2000), 87–90.

[28] D. Isvoranu, C. Udriste, *Fluid Flow Versus Geometric Dynamics*, BSG Proceedings 13, Geometry Balkan Press, 2006, pp. 70–82.

[29] P. Kim, *Invariantization of Numerical Schemes Using Moving Frames*, Springer, New York, 2007, pp. 525–546.

[30] L. Komzsik, *Applied Calculus of Variations for Engineers*, CRC Press Taylor & Francis Group, 2014.

[31] M. L. Krasnov, G. I. Makarenko, A. I. Kiselev, *Problems and Exercises in the Calculus of Variations*, Mir Publishers, Moscow, 1975.

[32] L. D. Landau, E. M. Lifshitz, *Mechanics*, Pergamon Press, Oxford, 1969.

[33] E. Langamann, *Introduction to Variational Calculus*, KTH Royal Institute of Technology, Stockholm, 2008.

[34] M. A. Lavrentiev, L. A. Liusternik, *Curs de Calcul Variaţional* (in Romanian), Editura Tehnică, 1955.

[35] B. Leimkuhler, S. Reich, *Simulating Hamiltonian Dynamics*, Cambridge University Press, Cambridge, 2005.

[36] P.-L. Lions, *Generalized Solutions of Hamilton-Jacobi Equations*, Pitman (Advanced Publishing Program), Boston, MA, 1982.

[37] P. D. Loewen: *On the Lavrentiev phenomenon*, Canadian Mathematical Bulletin 30, 1 (1987), 102–108.

[38] A. De Luca, G. Oriolo, *Modelling and control of nonholonomic mechanical systems*, in Kinematics and Dynamics of Multi-Body Systems (J. Angeles, A. Kecskemethy Eds.), Springer-Verlag, New York, 1995.

[39] A. De Luca, G. Oriolo, C. Samson, *Feedback control of a nonholonomic car-like robot*, in Robot Motion Planning and Control (J.-P. Laumond Ed.), Springer-Verlag, New York, 1998.

[40] B. Mania, *Sopra un esempio di Lavrentieff*, Bollettino dell'Unione Matematica Italiana 13 (1934), 147–153.

[41] J. E. Marsden, M. West, *Discrete mechanics and variational integrators*, Acta Numerica 10 (2001), 357–514.

[42] J. H. McClellan, R. W. Schafer, M. A. Yoder, *Signal Processing First*, Prentice Hall, Upper Saddle River, NJ, 2003.

[43] G. Mircea, M. Neamtu, D. Opris, *Deterministic, Uncertain, Stochastic and Fractional Differential Equations with Delay*, LAP LAMBERT Academic Publishing, Saarbrucken, Germany, 2011.

[44] T. C. A. Molteno, N. B. Tufillaro, *An experimental investigation into the dynamics of a string*, American Journal of Physics 72, 9 (2004), 1157–1169.

[45] O. M. Moreschi, G. Castellano, *Geometric approach to non-holonomic problems satisfying Hamilton's principle*, Revista de la Union Matematica Argentina, 47, 2 (2006), 125–135.

[46] R. M. Murray, Z. Li, S. S. Sastry, *A Mathematical Introduction to Robotic Manipulation*, CRC Press, Boca Raton, FL, 1994.

[47] M. Neamtu, D. Opris, *Economic Games. Discrete Economic Dynamics. Applications* (in Romanian), MIRTON Editorial House, Timisoara, 2008.

[48] P. J. Olver, *The Calculus of Variations*, 2021, http://www.math.umn.edu/~olver.

[49] D. Opris, C. Giulvezan, *Geometric Integrators and Applications* (in Romanian), Mirton Editorial House, Timisoara, 1999.

[50] G. Oriolo, *Control of Nonholonomic Systems*, Dottorato di Ricerca in Ingegneria dei Sistemi, DIS, Università di Roma "La Sapienza", 2008.

[51] L. A. Pars, *A Treatise of Analytical Dynamics*, Heinemann, London, 1965.

[52] I. N. Popescu, *Gravitation*, Editrice Nagard, Roma, Italy, 1988.

[53] M. Popescu, P. Popescu, H. Ramos, *Some new discretizations of the Euler–Lagrange equation*, Communications in Nonlinear Science and Numerical Simulation, 103, December (2021), 106002.

[54] R. Portugal, L. Golebiowski, D. Frenkel, *Oscillation of membranes using computer algebra*, American Journal of Physics 67, 6 (1999), 534–537.

[55] M. A. Rincon, R. D. Rodrigues, *Numerical solution for the model of vibrating elastic membrane with moving boundary*, Communications in Nonlinear Science and Numerical Simulation 12, 6 (2007), 1089–1100.

[56] D. M. Rote, C. Yigang, *Review of dynamic stability of repulsive-force maglev suspension systems*, IEEE Transactions on Magnetics, 38, 2 (2002), 1383–1390.

[57] I. B. Russak, *Calculus of Variations*, MA4311, Lectures Notes, Department of Mathematics, Naval Postgraduate School, Monterey, CA, 93943, July, 2002.

[58] G. E. Silov, *Analiză Matematică* (in Romanian), Editura Științifică și Enciclopedică, București, 1989.

[59] C. J. Silva, D. F. M. Torres, *On the Classical Newton's Problem of Minimal Resistance*, 3rd Junior European Meeting, University of Aveiro 6–8/September/2004.

[60] J. C. Slater, N. H. Frank, *Mechanics*, McGraw-Hill, New York, 1947.

[61] F. Suryawan, J. De Doná, M. Seron, *Methods for trajectory generation in magnetic levitation system under constraints*, Control & Automation (MED), 2010, 18-th Mediterranean Conference on ..., DOI: 10.1109/MED.2010.5547748

[62] N. B. Tufillaro, *Nonlinear and chaotic string vibrations*, American Journal of Physics 57, 5 (1989), 408–414.

[63] C. Udriste, *Field Lines*, Technical Publishing House, Bucharest, 1988 (in Romanian).

[64] C. Udriste, *Geometric Dynamics*, Mathematics and Its Applications, 513, Kluwer Academic Publishers, Dordrecht, Boston, London, 2000.

[65] C. Udriste, *Tools of geometric dynamics*, Buletinul Institutului de Geodinamică, Academia Română, 14, 4 (2003), 1–26; Proceedings of the XVIII Workshop on Hadronic Mechanics, honoring the 70th birthday of Prof. R. M. Santilli, the originator of hadronic mechanics, University of Karlstad, Sweden, June 20–22, 2005; Eds. Valer Dvoeglazov, Tepper L. Gill, Peter Rowland, Erick Trell, Horst E. Wilhelm, Hadronic Press, Palm Harbor, FL, 2006, pp. 1001–1041.

[66] C. Udriste, A. Udriste, *From flows and metrics to dynamics and winds*, Bulletin of the Calcutta Mathematical Society, 98, 5 (2006), 389–394.

[67] C. Udriste, D. Zugravescu, F. Munteanu, *Nonclassical electromagnetic dynamics*, WSEAS Transactions on Mathematics, 7, 1 (2008), 31–39.

[68] C. Udriste, L. Matei, *Lagrange-Hamilton Theories* (in Romanian), Monographs and Textbooks 8, Geometry Balkan Press, Bucharest, 2008.

[69] C. Udriste, M. Ferrara, D. Opris, *Economic Geometric Dynamics*, Geometry Balkan Press, Bucharest, 2004.

[70] C. Udriste, V. Arsinte, C. Cipu, *Von Neumann analysis of linearized discrete Tzitzeica PDE*, Balkan Journal of Geometry and Its Applications, 15, 2 (2010), 100–112.

[71] C. Udriste, I. Tevy, *Dynamical Systems and Differential Geometry via MAPLE*, Cambridge Scholars Publishing, Newcastle, 2021.

[72] C. Udriste, O. Dogaru, I. Tevy, *Extrema with Nonholonomic Constraints*, Monographs and Textbooks 4, Geometry Balkan Press, Bucharest, 2002.

[73] C. Udriste, O. Dogaru, M. Ferrara, I. Tevy, *Extrema with constraints on points and/or velocities*, Balkan Journal of Geometry and Its Applications, 8, 1 (2003), 115–123.

[74] P. J. van der Houwen, B. P. Sommeijer, *Approximate factorization for time-dependent partial differential equations*, Journal of Computational and Applied Mathematics, 128, 1–2 (2001), 447–466.

[75] B. van Brunt, *The Calculus of Variations*, Springer-Verlag, New York, 2003.

[76] A. P. Veselov, *Integrable discrete-time systems and difference operators*, Functional Analysis and Its Applications, 22, 2 (1988), 83–93.

[77] A. P. Veselov, *Integrable maps*, Uspekhi Mat Nauk, 46, 5 (1991), 3–45.

[78] G. Vranceanu, *Lectures of Differential Geometry* (in Romanian), Didactical and Pedagogical Editorial House, Bucharest, Vol. I, 1962; Vol. II, 1964.

[79] L. C. Young, *Lectures on the Calculus of Variations and Optimal Control Theory*, Chelsea, New York, 1980.

[80] R. Weinstock, *Calculus of Variations: with Applications to Physics and Engineering*, Revised Edition, Dover Publications, Inc., New York, 2012.

[81] L. K. Weiwei, Y. Tong, E. Kanso, J. E. Marsden, P. Schröder, M. Desbrun, *Geometric, Variational Integrators for Computer Animation*, Eurographics/ACM SIGGRAPH Symposium on Computer Animation, 2006; M.-P. Cani, J. O'Brien (Editors).

[82] J. M. Wendlandt, J. E. Marsden, *Mechanical integrators derived from a discrete variational principle*, Physica D: Nonlinear Phenomena, 106, 34 (1997), 223–246.

[83] E. T. Whittaker, *A Treatise on The Analytical Dynamics of Particles & Rigid Bodies*, Cambridge University Press, Cambridge, 1989.

Index

Variational Calculus with Engineering Applications, First Edition. Constantin Udriste and Ionel Tevy.
© 2023 John Wiley & Sons Ltd. Published 2023 by John Wiley & Sons Ltd.